DIGITAL
ELECTRONICS

DIGITAL ELECTRONICS

Ray Ryan and Lisa A. Doyle

With Contributions By
Abraham I. Pallas, Ph.D.
Columbia State Community College
Columbia, Tennessee

GLENCOE

Macmillan/McGraw-Hill

New York, New York   Columbus, Ohio   Mission Hills, California   Peoria, Illinois

As part of Glencoe's effort to help make a difference in the environment, this book has been printed on recycled paper.

Library of Congress Cataloging-in-Publication Data

Ryan, Ray.
   Digital electronics / Ray Ryan and Lisa A Doyle; with contributions by Abraham I. Pallas.
       p.   cm. — (Glencoe tech series)
   Rev. ed. of: Basic digital electronics. c1990.
   Includes index.
   ISBN 0-02-801306-9 (softcover)
   1. Digital electronics.   I. Doyle, Lisa A.   II. Pallas, Abraham.
III. Ryan, Ray.   Basic digital electronics.   IV. Title.   V. Series.
TK7868.D5R89   1993
621.381—dc20   92-34443
                                               CIP

Digital Electronics

Send all inquiries to:
GLENCOE DIVISION
Macmillan/McGraw-Hill
936 Eastwind Drive
Westerville, OH 43081

ISBN 0-02-801306-9

Printed in the United States of America.

1 2 3 4 5 6 7 8 9   MAL   00 01 99 98 97 96 95 94 93 92

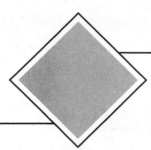

CONTENTS

4 Logic Gates   45

5 Logic Families   63

6 Boolean Algebra   85

7 *Combinational and Sequential Logic*   *107*

8 *Typical Logic Networks*   *139*

9 *Complex Logic Networks*   *163*

EDITOR'S FOREWORD

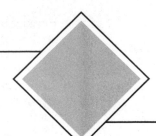

The Glencoe Tech Series is an economically packaged sequence of books tailored for use in an educational setting. The series is designed to provide the student with graduated learning experiences, ranging from dc and ac fundamentals through advanced electronics devices and integrated circuits, often with a minimum of mathematics. Each book contains chapter objectives, summaries, questions, and problems, all designed to enhance the student's learning. An instructor's manual with answers to questions and solutions to even-numbered problems is also available.

This text provides the student with an introduction to digital electronics. Beginning with basic digital concepts, the student progresses to combinatorial and sequential logic and concludes with basic microprocessor concepts. *Digital Electronics* is appropriate for a one-semester course in digital fundamentals, and it prepares the student to progress to the study of microprocessors. The student should have a knowledge of elementary algebra in order to understand and perform binary calculations.

Abraham I. Pallas
Project Editor

GLENCOE TECH SERIES
Books in this series:
Basic Electricity and Electronics by Delton T. Horn
DC/AC Electricity by Victor F. Veley
Electronic Devices by Joseph J. Carr
Digital Electronics by Ray Ryan and Lisa A. Doyle
Electronic Power Control by Irving M. Gottlieb

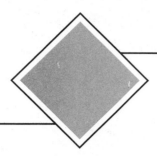

INTRODUCTION

*Y*ou can find digital circuits in every phase of electronics today. They are in such diverse applications as TV receivers, stereo circuits, telephones, in automobiles, and of course in computers. And even though digital circuits inherently operate at low voltages (one of their advantages), they are even performing in an increasingly larger portion of the various display and control functions in radio-frequency and high-power applications.

This book is for those who are familiar with basic electronics but don't have a firm grasp on of all the concepts of digital circuits. This text introduces the 1s and 0s of the digital world in a simplified manner so that any hobbyist, technician, or engineer can understand the principles involved without having to resort to special training.

The main emphasis throughout the book is on presenting the various types of circuits and describing the pertinent features of each. As is further explained in Chapter 1, today's electronics seldom use *discrete,* or individual, components in making up a circuit. Several components are *integrated* into a compact package that can often perform multiple functions. Therefore, there is a slight bias in this text that a given circuit performs a *function* rather than putting a lot of emphasis on specific circuit details. Ultimately, the reader should be concerned with the circuit's application, the concepts of which are explained in the later chapters.

The first several chapters describe the number systems and the different kinds of codes that digital systems use. The next several chapters explain the basics of *logic* and introduce the *logic diagram* and *truth table.* With this background, you can analyze and understand any digital circuit.

The remaining chapters are basically descriptions of applications of digital circuits. Starting with very simple functions made up of the basic *gates,* the circuits become increasingly more sophisticated until, finally, memories are described that consist of literally thousands of discrete components and perform a multitude of tasks with very high reliability. It is hoped this book fulfills a need for an easily understood guide and reference to digital concepts and circuitry.

1 ⟩ NUMBER SYSTEMS

AFTER YOU COMPLETE this chapter, you will be able to:

☐ Explain the different types of number systems used in digital systems
☐ Convert numbers among the number systems
☐ Convert a decimal fraction to a binary fraction
☐ Describe the three major signed binary number notations

*T*raditionally, analog circuits made up the bulk of all electronic circuitry. Recall the first computer, the ENIAC, which consisted of 18,000 vacuum tubes. This primitive computing machine weighed 30 tons and consumed an enormous amount of power. Today's computers are made up almost entirely of tiny digital components that can operate at increasingly faster speeds with increasingly less power.

The development of the transistor has practically replaced vacuum tubes in most applications (except where substantially high power is required). The transistor can be used as either an analog or digital device, but the trend is toward integrating hundreds—or thousands—of transistors and similarly operating devices into *digital integrated circuits* (ICs) that have capabilities far beyond those of discrete analog transistor circuits, especially in the areas of speed and efficiency.

Hence, although analog circuitry is still used for special applications that digital circuitry has not caught up to yet, digital electronics are quickly taking over most of the trends in electrical design. Analog circuits (especially vacuum tubes) are often required for high-power applications, but digital circuits provide the best options in speed, efficiency, accuracy, size, and cost.

THE CONCEPT OF DIGITAL

Consider the voltage level at the output of a transistor. Assume the input voltage is 12 volts. Biased as an analog circuit, the output could vary from 0 to 12 volts or any value between—a theoretically infinite number of possible values. However, say the transistor was wired as a *switch*. The device would then have only two conduction states: conducting or not conducting. The transistor is said to be either "on" or "off." In digital circuitry, these are the only two possible states at which a given point can exist.

These two states also can be expressed as "high" or "low," "true" or "false," or *1* or *0,* hence the term *digital.* Therefore, a digital circuit is one that expresses a voltage or current value as digits—1's or 0's. In many ways, this concept is rather simple to understand and manipulate compared to analog circuits.

The numbering concept that uses only the digits 1 and 0 is the *binary number system.* The first step is to understand the binary and other number systems and how to manipulate and convert them. Therefore, this chapter concentrates on the three basic number systems: decimal, binary, and octal.

DECIMAL NUMBER SYSTEM

The most common and familiar number system is the decimal number system. Decimal numbers are represented with the digits 0, 1, 2, 3, 4, 5, 6, 7, 8, and 9. In a multidigit decimal number, each position has a value that is 10 times the value of the next position to its immediate right. In other words, every position can be expressed by 10 raised to some power. The unit, or "ones" position is 10^0 (any number raised to the zero power, except zero, is equal to one). The tens position is 10^1, and the hundreds position is 10^2. A progression of increasing exponents can be continued to infinity to the left of the decimal point. The same progression also can be extended to the right of the decimal point, but the exponents are negative.

For example, the first position to the right of the decimal point is the tenths position; it has a weight of 10^{-1}. Thus, the form of every decimal number is

$$\dots 10^4 \; 10^3 \; 10^2 \; 10^1 \; 10^0 \, . \, 10^{-1} \; 10^{-2} \dots$$

Note that, in the decimal system, the most significant digit (the one with the largest power of 10) is always at the left, and the least significant digit is at the right. This convention is used throughout this book when writing numbers in any number system.

Other number systems can be represented in the same way. In each case, the weights of the various positions are powers of a particular number. This number is the *base,* or *radix,* of that number system and equals the number of digits used in that system (e.g., decimal = 10, binary = 2, octal = 8, etc.). Using the letter R to represent the radix of any number system, the form of any number is:

$$\ldots R^4\ F^3\ R^2\ R^1\ R^0\ .\ R^{-1}\ R^{-2}\ R^{-3} \ldots$$

If there is doubt as to the number system being employed, it should be clarified by writing the radix of the number as a subscript to the number. For example, 235_{10} indicates that the number 235 is being expressed in the decimal number system.

BINARY NUMBER SYSTEM

The binary number system is the most useful in digital circuits because there are only two digits (0 and 1). These *binary digits,* or *bits* as they are commonly called, are used to represent the states of switches, vacuum tubes, relays, transistors, etc.

Because the binary number system uses two digits, it has a radix of two, and the position weights are powers of two. The progression of binary numbers is:

$2^0 = 1$
$2^1 = 2$
$2^2 = 4$
$2^3 = 8$
$2^4 = 16$

.

.

.

Powers of two are listed in TABLE 1–1 for reference when dealing with binary numbers.

Table 1-1. Powers of Two.

n	2^n	n	2^n
1	2	41	219 90232 55552
2	4	42	439 80465 11104
3	8	43	879 60930 22208
4	16	44	1759 21860 44416
5	32	45	3518 43720 88832
6	64	46	7036 87441 77664
7	128	47	14073 74883 55328
8	256	48	28147 49767 10656
9	512	49	56296 99534 21312
10	1024	50	1 12589 99068 42624
11	2048	51	2 25179 98136 85248
12	4096	52	4 50359 96273 70496
13	8192	53	9 00719 92547 40992
14	16384	54	18 01439 85094 81984

Table 1-1.   Continued.

n	2^n	*n*	2^n
15	32768	55	36 02879 70189 63968
16	65536	56	72 05759 40379 27936
17	1 31072	57	144 11518 80758 55872
18	2 62144	58	288 23037 61517 11744
19	5 24288	59	576 46075 23034 23488
20	10 48576	60	1152 92150 46068 46976
21	20 97152	61	2305 84300 92136 93952
22	41 94304	62	4611 68601 84273 87904
23	83 88608	63	9223 37203 68547 75808
24	167 77216	64	18446 74407 37095 51616
25	335 54432	65	36893 48814 74191 03232
26	671 08864	66	73786 97629 48382 06464
27	1342 17728	67	1 47573 95258 96764 12928
28	2684 35456	68	2 95147 90517 93528 25856
29	5368 70912	69	5 90295 81035 87056 51712
30	10737 41824	70	11 80591 62071 74113 03424
31	21474 83648	71	23 61183 24143 48226 06848
32	42949 67296	72	47 22366 48286 96452 13696
33	85899 34592	73	94 44732 96573 92904 27392
34	1 71798 69184	74	188 89465 93147 85808 54784
35	3 43597 38368	75	377 78931 86295 71617 09568
36	6 87194 76736	76	755 57863 72591 43234 19136
37	13 74389 53472	77	1511 15727 45182 86468 38272
38	27 48779 06944	78	3022 31454 90365 72936 76544
39	54 97558 13888	79	6044 62909 80731 45873 53088
40	109 95116 27776	80	12089 25819 61462 91747 06176

n	2^n
81	24178 51639 22925 83494 12352
82	48357 03278 45851 66988 24704
83	96714 06556 91703 33976 49408
84	1 93428 13113 83406 67952 98816
85	3 86856 26227 66813 35905 97632
86	7 73712 52455 33626 71811 95264
87	15 47425 04910 67253 43623 90528
88	30 94850 09821 34506 87247 81056
89	61 89700 19642 69013 74495 62112
90	123 79400 39285 38027 48991 24224
91	247 58800 78570 76054 97982 48448

Counting in the binary number system is performed, much the same way as in the decimal number system. Recall that with decimal numbers, the count proceeds 0, 1, ... 9, until a single digit can no longer represent a number. Then a second digit is added and the count continues: 10, 11,

... 20, ... etc. Adding of digits is continued for higher and higher counts so that there are always enough digits to represent any number. The binary counts progress in a similar manner. Counting from zero to four in binary, the counts are

		Binary			Decimal
			0	=	0
			1	=	1
		1	0	=	2
		1	1	=	3
	1	0	0	=	4
Power of 2		2^2	2^1	2^0	

Note that each time the two digits 1 and 0 in one position are exhausted (counted as high as they will go), a 1 is added at the left, all digits to the right are made 0, and the count continues. The column to the far right represents the "ones" column, or 2^0 ($2^0 = 1$). The next columns to the left equal 2^1 ($2^1 = 2$) and 2^2 ($2^2 = 4$) respectively, as explained previously.

Expressing Binary Numbers

Although it is not a common practice to express leading zeros in the decimal system (i.e., writing 026, out of a possible 100, instead of just 26), such

Table 1-2.　Four-Digit Binary Count Sequence.

Decimal Number	Binary Number			
0	0	0	0	0
1	0	0	0	1
2	0	0	1	0
3	0	0	1	1
4	0	1	0	0
5	0	1	0	1
6	0	1	1	0
7	0	1	1	1
8	1	0	0	0
9	1	0	0	1
10	1	0	1	0
11	1	0	1	1
12	1	1	0	0
13	1	1	0	1
14	1	1	1	0
15	1	1	1	1
Power of Two:	2^3	2^2	2^1	2^0

zeros are included in binary counting for symmetry, convenience, and a view of the bigger picture, if applicable. Table 1–2 shows the decimal numbers 1 through 15 and each's binary equivalent. The maximum number of places required for the binary numbers in this case, is four, so that is the number of digits represented for each.

Binary-to-Decimal Conversion

Binary numbers can be converted to decimal form quite easily. The method for performing this conversion consists of adding together all of the position weights where a one appears. All positions containing a zero can be ignored.

The entire procedure can be thought of as a two-step process. First, determine the position weights by referring to TABLE 1–1 and write the proper position weights above the binary number to be converted. Second, add the position weight containing ones to obtain the resultant decimal number. Examples of binary-to-decimal conversions are

Powers of Two 2^5 2^4 2^3 2^2 2^1 2^0
Decimal Weights 32   16   8   4   2   1

Example 1:   1   1   0   1   0   1 = 32 + 16 + 4 + 1 = 53
Example 2:   1   0   1   0   0   1 = 32 + 8 + 1 = 41

Note that long binary expressions are broken up into groups of bits to aid in readability. These groups are called *bytes*. A byte is defined as a sequence of adjacent bits treated as a unit. Bytes are commonly made up of four, eight, or more bits, but greater numbers of bytes that are grouped together usually then form a *word*.

Decimal-to-Binary Conversion

Decimal-to-binary conversion is a more lengthy process. Also, there is more than one technique that can be used to perform the conversion. One method is the repeated subtraction of powers of two until either there is no remainder or the remainder that is left is sufficiently small for the desired accuracy of the conversion. This method works best for whole numbers, primarily because the positive powers of two can be remembered easily and because there are not too many digits to deal with.

Use the second method to convert a number containing a fraction (e.g., 1.25). Convert the whole part of the number first, then convert the decimal part. Convert the whole number part by successively dividing the number by two, (each time noting whether or not the number is exactly divisible by two). Each time there is a remainder of zero, write down a zero; and each time a one remains, write down a one. Convert the fractional part by repeatedly multiplying by two until the decimal portion of the number becomes exactly zero or until sufficient accuracy is obtained.

This method works well for fractional numbers because you need not know the negative powers of two to perform the conversion. Because the two methods differ significantly, each is discussed separately.

Subtraction Method To perform a decimal-to-binary conversion using the subtraction method, simply write down the powers of two up to and including the one closest to the number being converted. Then sequentially subtract each of these powers of two from the decimal number. If the power can be subtracted—that is, if it's not larger than the decimal remainder—then write a binary 1 in the bit position corresponding to that power of two. If the power is larger than the decimal remainder, write a 0 in that position and try the next lower power of two.

For example, consider the conversion for the decimal number 50. The largest power of two that is still smaller than 50 is $2^5 = 32$. Write down the powers of two from 2^5 down. The resultant subtractions are shown as:

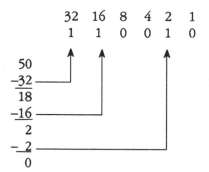

Two-Part Method First convert the whole part of the decimal number by repeatedly dividing by two. Each time there is no "remainder" (it divides evenly or is even), write down a 0. If there is a remainder (answer is odd), write a 1. The first division produces the least significant digit (the bit that represents the value smallest in magnitude) and the last division produces the most significant digit. A conversion for the number 53 is shown as:

Division	Remainder	
$53 \div 2 = 26$	1	least significant digit (LSB)
$26 \div 2 = 13$	0	
$13 \div 2 = 6$	1	
$6 \div 2 = 3$	0	
$3 \div 2 = 1$	1	
$1 \div 2 = 0$	1	most significant digit (MSB)

Result: 53_{10} = 1 1 0 1 0 1$_2$

Note that this method does not require writing down the powers of two.

Convert the fractional part by repeatedly multiplying by two. If the multiplication produces a whole number equal to one plus some fractional part, write a one. The multiplications by two are repeated until the fractional part exactly equals zero or until a sufficient number of digits have been obtained.

An example of the conversion of the decimal number 0.3125 follows. As before, note that it is not necessary to know the powers of two to perform the conversion.

Multiplication	Whole Number Part	
0.3125		
× 2		
0.6250	0	(MSB)
× 2		
1.2500	1	
× 2		
0.5000	0	
× 2		
1.0000	1	(LSB)

Result: 0.3125_{10} = 0 . 0 1 0 1$_2$

Signed Binary Numbers

Now that you understand the concept of a simple binary number, consider the representation of positive and negative numbers. Binary numbers that carry identification as to their polarity are referred to as *signed* binary numbers. Of practical interest is a method by which the plus and minus signs for positive and negative numbers can be represented in a digital format. Further, if a binary number is negative, there are several convenient ways of representing that number. Each representation has its own features. The three major signed binary number notations are: *sign magnitude notation, one's complement notation,* and *two's complement notation.*

Sign Magnitude   The most straightforward way to identify a signed binary number is to add a 0 or a 1 to the most significant bit of the overall number. This notation is called *sign magnitude* because the sign bit is given first, and then the positive magnitude of the number follows. A sign bit of zero indicates the number is positive, while a negative number is indicated by a sign bit of one. All other bits of the number indicate the magnitude of the number just as for an unsigned binary number.

Sign-magnitude notation is very easy to read, but it is not easy to use when adding and subtracting binary numbers. The decimal number 13 as a positive and a negative sign magnitude number is:

```
Sign
 0   1 1 0 1  = +13
 1   1 1 0 1  = -13
```

One's Complement   Another simple way to show a negative number is to attach a sign bit (as with the sign-magnitude notation) and invert all of the bits if the number is negative. This number representation is called *one's complement notation.* One's complement numbers are easy to form (invert all bits), but as shown in TABLE 1–3, there are two representations for zero. Also, when one's complement numbers are added and subtracted, a process, called *end-around carry* is necessary to obtain the correct answer. The decimal number 13 is shown below to indicate the difference between positive and negative numbers in one's complement.

```
Sign
 0   1 1 0 1  =  +13
 1   0 0 1 0  =  -13
```

Two's Complement   The most common representation for signed binary numbers is two's complement notation. The two's complement is generated by inverting the bits, as in one's complement, then adding one to the least significant bit (LSB). Two's complement is a little more difficult to generate, but it simplifies addition and subtraction. Further, there is only one representation for zero in two's complement.

The decimal number 13 is shown in two's complement notation to show the difference between positive and negative representations, as follows. (Refer to TABLE 1–3 for a comparison of two's complement numbers and other signed binary number notations.)

Table 1-3. Comparison of Signed Binary Numbers.

Decimal	Sign Magnitude	One's Complement	Two's Complement
15	0 1111	0 1111	0 1111
14	0 1110	0 1110	0 1110
13	0 1101	0 1101	0 1101
12	0 1100	0 1100	0 1100
11	0 1011	0 1011	0 1011
10	0 1010	0 1010	0 1010
9	0 1001	0 1001	0 1001
8	0 1000	0 1000	0 1000
7	0 0111	0 0111	0 0111
6	0 0110	0 0110	0 0110
5	0 0101	0 0101	0 0101
4	0 0100	0 0100	0 0100
3	0 0011	0 0011	0 0011
2	0 0010	0 0010	0 0010
1	0 0001	0 0001	0 0001
+0	0 0000	0 0000	0 0000
−0		1 1111	
−1	1 0001	1 1110	1 1111
−2	1 0010	1 1101	1 1110
−3	1 0011	1 1100	1 1101
−4	1 0100	1 1011	1 1100
−5	1 0101	1 1010	1 1011
−6	1 0110	1 1001	1 1010
−7	1 0111	1 1000	1 1001
−8	1 1000	1 0111	1 1000
−9	1 1001	1 0110	1 0111
−10	1 1010	1 0101	1 0110
−11	1 1011	1 0100	1 0101
−12	1 1100	1 0011	1 0100
−13	1 1101	1 0010	1 0011
−14	1 1110	1 0001	1 0010
−15	1 1111	1 0000	1 0001

Sign
```
0   1 1 0 1  =  +13
1   0 0 1 1  =  −13
```

OCTAL NUMBER SYSTEM

The octal number system has a radix of eight, so it uses eight digits (0,1,2,3,4,5,6,7). The position weights in the system are powers of eight, which also happen to be every third power of two. Thus TABLE 1–1 can also be used as a powers-of-eight table. For reference, the first six powers of eight are:

$$8^0 = 1$$
$$8^1 = 8$$
$$8^2 = 64$$
$$8^3 = 512$$
$$8^4 = 4096$$
$$8^5 = 32768$$

The octal number system is frequently used in digital circuits because it can easily be converted to binary. Also, there are significantly fewer digits in any given octal number than in the corresponding binary number, so it is much easier to work with the shorter octal numbers.

Binary-to-Octal and Octal-to-Binary Conversions

The conversion of a binary to an octal number or an octal to a binary number is simple since eight is the third power of two, providing a direct correlation between three-bit groups in a binary number and the octal digits. That is, each three-bit group of binary bits can be represented by one octal digit. The relationship of octal to binary digits is shown in TABLE 1-4.

Table 1-4.   Octal and Binary Equivalents.

Octal Digit	Binary Bits
0	000
1	001
2	010
3	011
4	100
5	101
6	110
7	111

Decimal-to-Octal Conversion

To convert a decimal number to octal, the simplest method is by division. The technique is much like the method for decimal-to-binary conversion. Begin by writing down the number to be converted; then repeatedly divide by eight. At each step of the division, any remainder represents the octal digit for one particular weight position. The conversion of decimal 91 to octal is:

$$
\begin{array}{ll}
\text{Division} & \text{Remainder} \\
91 \div 8 = 11 & 3 \quad \text{(LSB)} \\
11 \div 8 = 1 & 3 \\
1 \div 8 = 0 & 1 \quad \text{(MSB)}
\end{array}
$$

Result: $91_{10} = 133_8 = 1011011_2$

Octal-to-Decimal Conversion

An octal-to-decimal conversion can be done in the same manner as a binary-to-decimal conversion; that is, simply add up the position weights to obtain the decimal number. Hence, to reconvert 133_8 back to 91_{10}; the procedure is

$$
\begin{array}{lccc}
\text{Position Weights} & 64 & 8 & 1 \\
\text{Octal number} & 1 & 3 & 3
\end{array}
$$

$(64 \times 1) + (8 \times 3) + (1 \times 3)$

$= 64 + 24 + 3$

$= 91$

Summary

In a digital system, only two states are used. These may be high or low, on or off, 1 or 0, or true or false. Because there are only two states, a digital system is also called a *binary,* or *base 2, system.* In the binary system, negative numbers are expressed in three different ways: sign magnitude notation, one's complement notation, and two's complement notation. Other numbering systems are *octal* (base 8) and *decimal* (base 10). Various methods are available to convert numbers among these different systems.

Questions

1. What is meant by a digital system? How does it differ from an analog system?

2. What are the two states in a binary system?

3. How are the different place values determined in a numbering system?

4. What are the three different ways in which signed binary numbers are represented?

5. What is meant by an octal numbering system? a decimal numbering system?

Problems

1. Convert each of the following binary numbers to decimal and to octal: **a.** 1101   **b.** 1001.11

2. Repeat Problem 1 with   **a.** 0111   **b.** 1110.011

3. Convert each of the following decimal numbers to binary and to octal: **a.** 29   **b.** 51.5

4. Repeat Problem 3 with   **a.** 37   **b.** 58.675

5. Convert each of the following octal numbers to binary and to decimal: **a.** 43   **b.** 73.5

6. Repeat Problem 5 with   **a.** 25   **b.** 62.3

7. Express each of the following decimal numbers in sign magnitude, one's complement, and two's complement form: **a.** 3   **b.** −15

8. Repeat Problem 7 with   **a.** 11.   **b.** −5

BINARY CODES

AFTER YOU COMPLETE this chapter, you will be able to:

☐ Describe BCD, 8421, Excess-Three, and hexadecimal coding
☐ Understand what is meant by a cyclic code and why it is used
☐ Understand error-detecting and error-correcting codes
☐ Describe what parity is and how it is used for error-detecting
☐ Distinguish between different types of alphanumeric codes

*N*umbers expressed in signed binary form as described in chapter 1 are excellent for computational purposes in the inner workings of a digital machine. However, when numbers arc to be displayed, a somewhat different format is required so that operators of the machine do not need to deal in the awkward binary number system. Furthermore, special codes are useful to convert analog quantities to digital form and also to detect and correct errors. Hence, a number of special-purpose codes have been developed, each suited to specific functional requirements. As you can imagine, there is an almost an unlimited number of variations on codes that can be formed for one purpose or another. This chapter deals only with the most common of these codes and indicates some typical applications for each.

BINARY-CODED DECIMAL

Circuits and machines can deal readily with binary numbers, but people are accustomed to working with decimals, and there are considerably fewer decimal digits required to represent a number than there are binary. It is much easier to remember just a few digits than it is to remember many. Thus, whenever there is an interface between digital circuits and people, the interface data usually takes the decimal form. As a result the digital circuits must utilize some binary code to conveniently represent the decimal numbers. Several such codes are commonly used, all of which come under the classification of *binary-coded decimal* (BCD) format. Typically, a binary-coded decimal contains four binary bits for each decimal digit.

8421 Code

One example of a BCD format is the 8421 code. With this code, the decimal number 479 would be represented with twelve bits as 0100 0111 1001. Although this number contains only ones and zeroes, it is not a true binary number because it does not follow the rule that each bit is weighted by an increasing power of two. Instead each four-bit grouping has a binary weight, but each group represents the single-digit decimal numbers 0 through 9.

For each four-bit combination, the least significant bit has a weight of one, the next bit has a weight of two, the next a weight of four, and the most significant bit has a weight of eight, thereby deriving its name. As can be seen, the 8421 code has a standard binary weighting for each four bits; therefore, it is the most common BCD code and is often referred to as simply BCD.

As an example, express the decimal number 479 in binary and BCD:

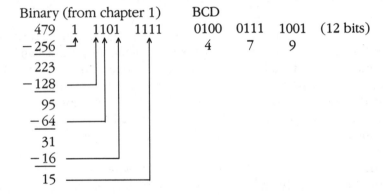

Therefore, BCD codes are less efficient than straight binary numbers because they need more bits to represent a given number. The number 479 required 12 bits in BCD but can be represented with 9 bits in pure binary. To be more specific, 4 binary bits can represent a maximum of 16 combinations, but a group of 4 bits in BCD only contains a maximum of 10 combinations.

Excess-Three Code

Whereas the 8421 code makes use of the first ten combinations of four bits, the *excess-three* code is a BCD format obtained by adding three to each decimal digit and coding the result as an ordinary binary number of four bits. A comparison of the 8421 code and excess-three code is provided in TABLE 2-1. In the table, the ten combinations of the excess-three code are "symmetrically" positioned within the 16 binary combinations so

**Table 2-1.   Comparisons of
BCD Codes.**

Binary	8421 Code	Excess-Three Code
0000	0000 = 0	
0001	0001 = 1	
0010	0010 = 2	
0011	0011 = 3	0011 = 0
0100	0100 = 4	0100 = 1
0101	0101 = 5	0101 = 2
0110	0110 = 6	0110 = 3
0111	0111 = 7	0111 = 4
1000	1000 = 8	1000 = 5
1001	1001 = 9	1001 = 6
1010		1010 = 7
1011		1011 = 8
1100		1100 = 9
1101		
1110		
1111		

that there are three unused states preceding and following the allowed combinations. This symmetry permits the excess-three code to be more easily manipulated arithmetically. Because this code does not use the all-zero state, this feature can be used as an error-detecting device. For example, if any excess-three number should occur with no ones in it, then some equipment malfunction might be causing an error. As with the 8421 code, there are exactly six "unused" combinations of the four binary bits.

Hexadecimal Code

It seems a shame to waste all those unused states for a BCD code, yet it is so much easier to segment the binary information into 4-bit groupings. A code that utilizes all 16 states for each group of 4 bits is the hexadecimal code. This code is used frequently by digital computer personnel where very large groupings of 8, 16, or 32 bits are the normal form of data transfer.

The hexadecimal code for the 16 combinations is shown in TABLE 2-2. One digit or letter represents each 4-bit combination. Consider a 16-bit piece of information a computer technician might analyze. The 16 bits for this example are read in pure binary as 1110011010100101. But coded into hexadecimal, the bits are regrouped and read as follows:

Binary	1110	0110	1010	0101
Hexadecimal	E	6	A	5

Clearly, E6A5 is much easier to remember than all those ones and zeroes.

Table 2-2.
Hexadecimal Code.

Binary	Hexadecimal Code
0000	0
0001	1
0010	2
0011	3
0100	4
0101	5
0110	6
0111	7
1000	8
1001	9
1010	A
1011	B
1100	C
1101	D
1110	E
1111	F

CYCLIC CODES

Another class of binary format is the *cyclic* coding system. These codes have the property of a single bit change when counting from one state to the next. Normally, a standard binary code changes several bits when going from one count to the next adjacent count. For example, when a binary number changes from 7(0111) to 8(1000), all four bits change simultaneously. A cyclic code would only change one of the bits, say from 0111 for 7 to 1111 for 8. Certainly this code does not have a binary weighting.

The property of a single-bit change is advantageous where the output of some analog device must be digitized. Usually, an electromechanical device such as a shaft encoder is used to perform the conversion. However, a shaft can stop at any random position—for example, on the border of changing from 0111 to 1000. If this situation occurs, each bit could be in any possible state, depending on the precision with which the electromechanical device was constructed. In this situation, the code being read out of the device is totally ambiguous. To eliminate this problem, a cyclic code is used. Because the code only changes one bit from position to position, the ambiguity at any position is at most one bit. The disadvantage to all cyclic codes is that arithmetic is extremely difficult to perform. If arithmetic manipulation is required, the cyclic code is usually converted to binary.

Gray Code

The most common cyclic format is called the *gray code*. Gray code is formed from pure binary by comparing each bit with its next adjacent bit, starting with the least significant bit and forming a *modulo two* sum, a binary sum with no carry, from the comparison. Thus, when two bits are compared, if they are alike (both ones or both zeroes) a gray code bit of zero will be generated. If the two bits are different, a gray code bit of one will be generated. For example,

$$57_{10} = \quad 1\ 1\ 1\ 0\ 0\ 1 \qquad \text{Binary}$$
$$0\ 1\ 1\ 1\ 0\ 0 \qquad \text{Shifted right one place}$$
$$1\ 0\ 0\ 1\ 0\ 1 \qquad \text{Modulo two sum} = \text{gray code}$$

If desired, gray code can be converted back to binary by a similar process, but starting with the most significant bit. If a one occurs, the next binary bit will be the opposite of the previous bit, but if a zero occurs, the next binary bit will be the same. In the example here, the first one encountered is copied directly, since it represents a change from the zero state to the one state. Thereafter, each one represents an additional change in state from the bit just written down and each zero represents no change in state.

1	0	0	1	0	1	Gray code
	No	No		No		
Copy	Change	Change	Change	Change	Change	
1	→ 1	→ 1	→ 0	→ 0	→ 1	Binary

A complete listing of the gray code equivalents for the first 16 binary combinations is shown in TABLE 2-3. Note that although several bits of the binary code might change between adjacent counts, only one bit changes for the gray code.

ERROR-DETECTING CODES

Modern digital equipment is complex and there is always a possibility of mechanical or electrical failure in some portion of the unit. For this reason, error detecting codes have been devised that can check the result of a data transmission or an arithmetic operation. These codes consist of in-

**Table 2-3.
Comparison of Gray
Code to
Binary Counts.**

Binary	Gray Code
0000	0000
0001	0001
0011	0010
0100	0110
0101	0111
0110	0101
0111	0100
1000	1100
1001	1101
1010	1111
1011	1110
1100	1010
1101	1011
1110	1001
1111	1000

formation bits plus some *check* bits. The check bits can be considered redundant, in that the resultant code contains more bits than are actually necessary to represent a particular piece of information.

The BCD codes previously described already contain limited amounts of error-detecting capability. Because BCD uses only 10 of 16 states for each four-bit group, the 6 states that do not occur represent "forbidden" combinations. If any of these combinations does occur, it is because of an error. Also, the excess-three code has an additional error-detecting feature: the all-zero state is one of the forbidden combinations. Because an all-zero state represents the absence of a signal, this is a useful fault-detecting feature in digital circuits.

The most common method of detecting errors is through the use of *parity bits.* Usually, a single parity bit is added to the information bits such that the number of ones in the data is odd or even, as desired. For example, suppose that the binary number 1001 were to be transmitted with a parity bit added, which would make the number of ones odd. Then, the actual binary data would consist of five bits, as shown:

$$1 \quad 0 \quad 0 \quad 1 \qquad 1$$
$$\text{data bits} \qquad \text{parity bit}$$

This scheme is called *odd parity* because the goal is that there be an odd number of ones in the data. The proper odd-parity bits for the first 16 binary combinations are shown in TABLE 2-4.

Table 2-4.   Odd Parity for Four Data Bits.

Decimal Number	Binary Number	Parity Bit
0	0000	1
1	0001	0
2	0010	0
3	0011	1
4	0100	0
5	0101	1
6	0110	1
7	0111	0
8	1000	0
9	1001	1
10	1010	1
11	1011	0
12	1100	1
13	1101	0
14	1110	0
15	1111	1

In a similar fashion, *even parity* can be used. Here, the object is to always have an even number of ones in the data. The parity bit for an even parity scheme is the opposite of that required to form odd parity.

A single parity bit can detect the existence of an error if only one error occurs or if the number of errors is odd. If an even number of errors occurs, they will go undetected by this scheme. Although there is no significant advantage to using odd or even parity, odd parity is often used because it detects the absence of any signal (the all-zero state).

ERROR-CORRECTING CODES

Parity checking, forbidden combinations of BCD codes, and the use of gray code are several of the methods that can detect when an error has occurred. Nevertheless, these methods do not determine what the error is or correct it. There is a class of codes, called *error-correcting codes,* that performs this function.

The *Hamming* code is the most widely used single-error correcting code. Basically, the Hamming code is a multiple parity scheme. For a given number of bits, there is an associated number of even parity bits. These parity bits check different groupings of the data bits for even parity, with each parity bit checking a separate grouping. An example of the Hamming code applied to a four-bit binary number is shown in TABLE 2-5. From the

Table 2-5.   Hamming Code for Four Data Bits.

Decimal Number	8	4	2	1	P1	P2	P3
0	0	0	0	0	0	0	0
1	0	0	0	1	1	1	1
2	0	0	1	0	0	1	1
3	0	0	1	1	1	0	0
4	0	1	0	0	1	0	1
5	0	1	0	1	0	1	0
6	0	1	1	0	1	1	0
7	0	1	1	1	0	0	1
8	1	0	0	0	1	1	0
9	1	0	0	1	0	0	1
10	1	0	1	0	1	0	1
11	1	0	1	1	0	1	0
12	1	1	0	0	0	1	1
13	1	1	0	1	1	0	0
14	1	1	1	0	0	0	0
15	1	1	1	1	1	1	1

table, it can be verified that the three parity bits—P1, P2, and P3—are associated with the data bits as follows: P1:8-4-1, P2:8-2-1, P3:4-2-1. Each parity bit represents even parity for its associated three data bits. Comparison of these three parity bits will not only indicate when there is an error, but will also indicate which bit is wrong.

The mechanism for determining and correcting errors lies in the regeneration of parity at the receiving end. Three new parity bits are generated for three four-bit groups that now include the transmitted parity bits. By regrouping the seven received bits as shown in TABLE 2-6, the result indicated by the formation of the new parity bits locates the bit position in which an error occurred. For example, consider an error in transmitting the number 13 (1010101) such that the code is received (1010111). The three new parity bits are computed as follows:

Received	P1	P2	8	P3	4	2	1
Code	1	0	1	0	1	1	1

New P1 for (P1:8-4-1) = 0
New P2 for (P2:8-2-1) = 1
New P3 for (P3:4-2-1) = 1

The binary number P3P2P1 = 110 indicates that an error occurred in bit six counting from the left. A correction (inversion) can now be performed on this bit. If the check number P3P2P1 = 000, then the given number is correct. Obviously, only one error can be corrected with this code, and it is thus labeled a single-error correcting code. If more check bits were added, it would be possible to correct more than one error.

**Table 2-6. Hamming Code Regrouped
for Error Correction.**

Decimal Number	*P1*	*P2*	*8*	*P3*	*4*	*2*	*1*
0	0	0	0	0	0	0	0
1	1	1	0	1	0	0	1
2	0	1	0	1	0	1	0
3	1	0	0	0	0	1	1
4	1	0	0	1	1	0	0
5	0	1	0	0	1	0	1
6	1	1	0	0	1	1	0
7	0	0	0	1	1	1	1
8	1	1	1	0	0	0	0
9	0	0	1	1	0	0	1
10	1	0	1	1	0	1	0
11	0	1	1	0	0	1	1
12	0	1	1	1	1	0	0
13	1	0	1	0	1	0	1
14	0	0	1	0	1	1	0
15	1	1	1	1	1	1	1

ALPHANUMERIC CODES

Codes that represent alphabetic characters and symbols, as well as numbers, are called *alphanumeric codes*. These codes can also be used to represent instructions for conveying information; for example, in higher-level computer languages.

At a minimum, an alphanumeric code must represent 10 digits and 26 letters of the alphabet, for a total of 36 items. This requires six bits in each code combination, since five bits are insufficient ($2^5 = 32$). There are 64 total combinations of six bits, so there are 28 unused code combinations. These remaining combinations can be used to function as punctuation, spaces, and other instructions for the receiving system. Two alphanumeric codes are ASCII and EBCDIC, discussed subsequently.

ASCII

One standardized alphanumeric code is the American Standard Code for Information Interchange (ASCII). This code is probably the most widely used type. It is a seven-bit code in which the decimal digits are represented by the 8421 BCD code preceded by 011. The letters of the alphabet and other symbols and instructions are shown in TABLE 2-7. For instance, the letter A is represented by the code 1000001, the comma by 0101100, and the command ETX (end of text) by 0000011.

EBCDIC

Another alphanumeric code frequently encountered is the Extended Binary-Coded Decimal Interchange Code (EBCDIC). This is an eight-bit code in which the decimal digits are represented by the 8421 code preceded by 1111. Both lowercase and uppercase letters can be represented with this code, in addition to numerous other symbols and commands.

Table 2-7. American Standard Code for Information Interchange.

	000	*001*	*010*	*011*	*100*	*101*	*110*	*111*
0000	NUL	DLE	SP	0	@	P	`	p
0001	SOH	DC$_1$	!	1	A	Q	a	q
0010	STX	DC$_2$	"	2	B	R	b	r
0011	ETX	DC$_3$	#	3	C	S	c	s
0100	EOT	DC$_4$	$	4	D	T	d	t
0101	ENQ	NAK	%	5	E	U	e	u
0110	ACK	SYN	&	6	F	V	f	v
0111	BEL	ETB	'	7	G	W	g	w
1000	BS	CAN	(	8	H	X	h	x
1001	HT	EM	)	9	I	Y	i	y
1010	LF	SUB	*	:	J	Z	j	z
1011	VI	ESC	+	;	K	[	k	{
1100	FF	FS	,	<	L	\	l		
1101	CR	GS	-	=	M	]	m	}
1110	SO	RS	.	>	N	^	n	~
1111	SI	US	/	?	O	__	o	DEL

Definitions of Control Abbreviations:

ACK	Acknowledge	FS	Form separator
BEL	Bell	GS	Group separator
BS	Backspace	HT	Horizontal tab
CAN	Cancel	LF	Line feed
CR	Carriage return	NAK	Negative acknowledge
DC$_1$–DC$_4$	Direct control	NUL	Null
DEL	Delete idle	RS	Record separator
DLE	Data link escape	SI	Shift in
EM	End of Medium	SO	Shift out
ENQ	Enquiry	SOH	Start of heading
EOT	End of transmission	STX	Start of text
ESC	Escape	SUB	Substitute
ETB	End of transmission block	SYN	Synchronous idle
ETX	End text	US	Unit separator
FF	Form feed	VT	Vertical tab

Summary

Binary codes require too many digits to be useful in displays. They are often replaced by one of a group of binary-coded decimal (BCD) codes. Examples of these codes are 8421 and excess-three codes. Hexadecimal (base 16) code is also commonly used. To help avoid errors when going from one binary number to the next, you may use a cyclic code. The most common of these is called *gray code*. One way to detect errors is to use a parity bit. Error correction can be done by using an error-correcting code. The most common of these is the Hamming code. There are two main codes for alphanumeric coding: ASCII and EBCDIC. They are used to code letters, numbers, and control information in binary form.

Questions

1. What is meant by BCD?

2. What are two different BCD codes? How are they different?

3. What is hexadecimal? How does it differ from BCD codes?

4. What is meant by a cyclic code? Why is it used? What is one example of a cyclic code?

5. Why are error-detecting codes used? What is the most common form of error detection?

6. What is meant by parity? How is it used?

7. What is an error-correcting code? What is one example of an error-correcting code?

8. What is the term used for codes that represent alphabetic characters and symbols? What are two examples of such a code?

Problems

1. Express the decimal number 135 in   **a.** binary   **b.** 8421
   c. excess-three   **d.** hexadecimal

2. Repeat Problem 1 for the decimal number 748.

3. Convert the binary number 1011 1100 to gray code.

4. Convert the binary number 0101 0110 to gray code.

5. If odd parity is being used, what is the parity bit when the decimal number 43 is converted to binary?

6. If even parity is being used, what is the parity bit when the decimal number 83 is converted to binary?

7. The decimal number 6 is to be transmitted using the Hamming error-correcting code.   **a.** What are the values of the parity bits P1P2P3?   **b.** What 7-digit binary number is transmitted?   **c.** If the binary number 1100111 is received, how can the location of the error be determined?

8. The binary number 1101111 is received, using the Hamming error-correcting code. Find the error in transmission.

9. Write the ASCII code for the word END.

10. Write the ASCII code for the word HELLO.

BINARY ARITHMETIC

AFTER YOU COMPLETE this chapter, you will be able to:

☐ Perform arithmetic operations using binary numbers
☐ Perform sign magnitude arithmetic operations
☐ Perform operations using one's and two's complement notation
☐ Describe 8421 and excess-three codes

The various binary number representations and binary codes all have unique characteristics that make them convenient in some particular application. For pure binary bits, there are very simple rules that always apply when adding, subtracting, multiplying, or dividing. But when bits are combined into one of the specialized notations described in chapters 1 and 2, care must be taken to recognize the proper notation and to treat these bits accordingly. For example, 10111 (which stands for 23 in pure binary) means -7 in sign magnitude notation, -8 in one's complement notation, -9 in two's complement notation, 17 in 8421 BCD, and 14 in excess-three BCD. Clearly, binary arithmetic operations must be varied to suit the particular number representation format or scheme being used.

BASIC RULES

The binary number system has the same basic format as any other number system, and thus, arithmetic operations are performed in the same manner as in other systems. For example, in the binary system, $0 + 0 = 0$, $0 + 1 = 1$, and $1 + 0 = 1$. However, because there is no single binary digit for the number 2, $1 + 1 = 0$ and there is a 1 to carry over. This carry is similar to a decimal carry when the decimal sum exceeds the number 9. Table 3-1 shows addition, subtraction, multiplication, and division examples for the binary number system.

Multiplication and division are performed in binary in the same manner as with decimal numbers. It involves forming the partial products, shifting each successive partial product left one place, then adding all the

Table 3-1.   Binary Arithmetic.

Binary Addition

$0 + 0 = 0$		
$0 + 1 = 1$	13	01101
$1 + 0 = 1$	$+14$	01110
$1 + 1 = 0$, carry 1	27	11011

Binary Subtraction

$0 - 0 = 0$		
$0 - 1 = 1$, borrow 1	27	11011
$1 - 0 = 1$	-11	01011
$1 - 1 = 0$	16	10000

Binary Multiplication

$0 \times 0 = 0$		10101
$0 \times 1 = 0$		101
$1 \times 0 = 0$		10101
$1 \times 1 = 1$	21	00000
	$\times 5$	10101
	105	1101001

Binary Division

$0 \div 0 = ?$	21	10101
(Undefined)	5) 105	101) 1101001
$0 \div 1 = 0$	10	101
$1 \div 0 = ?$	5	110
(Undefined)	5	101
$1 \div 1 = 1$		101
		101

partial products. The following examples illustrate the procedure and the equivalent decimal multiplication or division.

As the table shows, multiplication and division are generally the most complex arithmetic operations and require several steps to perform. But if the number to be used as a multiplier or divisor happens to be a power of two, the operation becomes extremely simple. To multiply by a power of two, shift the number being multiplied by a number of digits equal to the power of two it is being multiplied by. Consider the multiplication $13 \times 8 = 104$. The multiplication is:

$$13 \quad \times \quad 8 \quad = \quad 104$$

$$\lfloor 1\ 1\ 0\ 1 \rfloor \quad \times \quad 1\ 0\ 0\ 0 \quad = \lfloor 1\ 1\ 0\ 1 \rfloor\ 0\ 0\ 0 \rfloor$$

Original Number —————————————— Three-place left shift

Similarly, if a number happens to be divided into a power of two, a right shift by the proper number of digits will perform the division. Suppose, for example, that the decimal number 13 is to be divided by 4 (13/4 = 3.25) again noting that 4 is the second power of 2. The division is:

$$13 \quad \div \quad 4 \quad = \quad 3.25$$

$$\lfloor 1\ 1\ 0\ 1 \rfloor \quad \div \quad 1\ 0\ 0 \quad = \lfloor 0\ 0 \lfloor 1\ 1\ .\ 0\ 1 \rfloor$$

Original Number ———————— Two-place right shift

SIGN MAGNITUDE NOTATION

Addition and subtraction of a sign magnitude number follows the basic rules given for the binary number system but includes with each number a sign bit. The sign bit is actually another plus or minus sign, where plus is denoted by 0 and minus is denoted by 1. As in decimal arithmetic, extreme care must be exercised to make sure all signs are accounted for; for example, $(+7) + (-7) = 0$ in the decimal number system, and 0 0111 + 1 0111 = 0 0000 in sign magnitude notation.

An example of a sign magnitude addition follows. Note that due to the sign bits, a subtraction is actually performed just as a subtraction is performed for the decimal addition shown at the side. Also, note that two separate operations are required for sign magnitude notation: addition and subtraction. Both one's complement notation and two's complement notation require only one operation: addition. A sign magnitude addition example follows:

$$
\begin{array}{rcl}
+14 & 0 & 1110 \\
+(-7) & 1 & 0111 \\
\hline
+7 & 0 & 0111
\end{array}
$$

Multiplication of sign magnitude numbers is done exactly as in TABLE 3-1, using the sign bits only for determining the sign of the product. Because multiplication is a process involving a number of steps, an actual procedure used by digital machines is examined here to show how the individual operations are performed. Sign magnitude multiplication can be accomplished by repeatedly adding either the number being multiplied or zero, and shifting one place to the right after each operation. Consider the example shown in TABLE 3-2. The multiplier bits have been labeled B2, B1, and B0 for convenience. Bit B2 is the most significant bit of the multiplier and bit B0 is the least significant bit. Each time the multiplier bit is 1, 10101 is added and a right shift is performed. When the multiplier bit is 0, 00000 is added and a right shift is performed. At each step, an accumulated result is maintained and the process continues until all bits of the multiplier have been used as operators.

Table 3-2.   Example of Sign Magnitude Multiplication.
Problem: 10101 × 101 = ?

Multiplier	*Operation*	*Accumulative Result*
B2B1B0 = 101		
B0 = 1	Add 10101	00000 (Starting point = 0)
(LSB)		+ 10101
		10101
B1 = 0	Shift right	010101
	Do nothing	00000
	(Add 00000)	010101
B2 = 1	Shift right	0010101
(MSB)	Add 10101	10101
		1101001 (Final product)

Result: 10101 × 101 = 1101001
21 × 5 = 105
NOTE: A right shift is not required after the last addition.

Operation Codes
Multiplier bit = 0
Do nothing (Add 0)
Multiplier bit = 1 Add

At the final step, no further right shift is required. There is nothing special about the two numbers used in this example; indeed, the procedure described works with any two *sign magnitude* numbers, either number consisting of as many bits as desired. Since the sign bits are not used for the actual multiplication procedure, it is apparent that the method works equally well for a pure binary number.

A sign magnitude division uses repeated subtractions and left shifts to obtain its results. The division process is usually somewhat more complicated than multiplication, in that a division does not necessarily produce a specific number of digits. Just as one divided by three (in decimal) results in an endless string of threes ($\frac{1}{3}$ = 0.333333333 . . .), there are many binary fractions that produce a very large number of digits. Therefore, a trial-and-error procedure is required, in which repeated division operations are performed until the desired accuracy is obtained or until a remainder of zero occurs. The division is terminated when the desired number of digits is obtained. Any time a remainder of zero occurs, the division is complete and the resultant answer is exact.

Table 3-3 shows an example of a typical sign magnitude division. Initially, note that the divisor is smaller than the number being divided. For

Table 3-3.   Example of Sign Magnitude Division.
Problem: 1101001 ÷ 101 = ?

Operation	Accumulative Result	Result ≥ Divisor?	Quotient
	01101001 (starting point)	—	
Shift left	11010010	Yes	1 (MSB)
Subtract	− 10100000		
	00110010		
Shift left	01100100	No	0
Do nothing	− 00000000		
	01100100		
Shift left	11001000	Yes	1
Subtract	− 10100000		
	0010100		
Shift left	01010000	No	0
Do nothing	− 00000000		
	01010000		
Shift left	10100000	Yes	1 (LSB)
Subtract	− 10100000		
	00000000		

Tentative answer: 0.10101
Final answer: 10101
Result: 1101001 ÷ 101 = 10101

NOTES: At start of problem divisor must be greater than number being divided. If not, shift left until it is. At end of problem, shift answer same number of places.

Operation Codes

Result ≥ Divisor?	Operation	Quotient
Yes	Subtract	1
No	Do nothing (Subtract 0)	0

this type of binary division, it is required that the divisor be the larger number. Hence, a five-place shift is performed to increase the size of the divisor (remember this information for later use).

Begin the division process by writing down the number to be divided, in this case 1101001. For each step of the process, perform a left shift, then examine the new number to determine whether or not it is larger than the divisor. If it is, the quotient is a 1 for that bit and the divisor is subtracted from the accumulative result. If the new number is smaller than the divisor, the quotient is a 0 for that bit and 00000000 is subtracted from the accumulative result.

This shift and subtract procedure is repeated until enough digits have been generated for the desired quotient accuracy, or until an accumulative result of zero occurs. Write down the final quotient in the form 0.*xxxxxxxx*, with a left shift to be performed equal to the number of places of left shift used when starting the problem. In this example, the quotient is initially written 0.10101 then shifted left five places, making the final answer 10101.

ONE'S COMPLEMENT NOTATION

One's complement addition is just like straight binary addition, except that there is an added bit, the sign bit. The sign bit is included as one of the bits that must be added. When all bits are added in the normal binary fashion, a positive or negative one's complement number is obtained by including the carry bit as part of the addition. If the carry bit is a zero, the answer is correct and no further correction is required. If the carry bit is a one, a one must be added to the sum to obtain a correct answer. The process of adding the carry bit to the sum is called *end-around carry*. An example of one's complement addition is shown here:

$$
\begin{array}{rcccccc}
+\ 14 & & 0 & 1 & 1 & 1 & 0 \\
+\ (-7) & & +\ 1 & 1 & 0 & 0 & 0 \\
\hline
+\ 7 & \boxed{1} & 0 & 0 & 1 & 1 & 0 \\
& & & & & & \rightarrow 1 \\
\hline
& & 0 & 0 & 1 & 1 & 1 \\
\end{array}
$$

One of the main advantages to one's complement numbers is that subtraction is never required. If a subtraction is called for, the number being subtracted is simply inverted (complemented) then the two numbers are added. Exactly the same rules apply for the addition. Addition in place of subtraction is a great convenience, and if the end-around carry could also be eliminated, the arithmetic would be further simplified. Two's complement numbers, considered next, provide this feature. An example of subtraction of one's complement numbers follows:

$$
\begin{array}{rr}
+\ 10 \\
-\ (-5) \\
\hline
+\ 15
\end{array}
\qquad
\begin{array}{r}
0\ \ 1\ 0\ 1\ 0 \\
-\ 1\ \ 1\ 0\ 1\ 0 \\
\hline
\end{array}
\rightarrow
\begin{array}{r}
0\ \ 1\ 0\ 1\ 0 \\
+\ 0\ \ 0\ 1\ 0\ 1 \\
\hline
0\ \ 1\ 1\ 1\ 1
\end{array}
$$

TWO'S COMPLEMENT NOTATION

Two's complement notation is the most widely used notation where arithmetic operations are required. It is extremely easy to use for addition and subtraction, and it lends itself readily to the multiplication and division operations. The addition of two's complement numbers is just like one's complement addition, except that the end-around carry is not required. All bits including the sign bits are added and the result is the sum. Any carry is ignored. An example of two's complement addition is shown here:

$$
\begin{array}{rr}
+\ 14 \\
+\ (-\ 7) \\
\hline
+\ 7
\end{array}
\qquad
\begin{array}{r}
0\ \ 1\ 1\ 1\ 0 \\
+\ 1\ \ 1\ 0\ 0\ 1 \\
\hline
0\ \ 0\ 1\ 1\ 1
\end{array}
$$

Two's complement numbers, like one's complement numbers, are subtracted by complementing, then adding. Remember that two's complements are formed by inverting and adding 1. Thus, to subtract any two numbers, invert the number being subtracted, add the two numbers, then add 1. When adding the two numbers, use the same rules as for the two's complement addition just described. An example of two's complement subtraction is shown here:

$$
\begin{array}{rr}
+\ 10 \\
-\ (-\ 5) \\
\hline
+\ 15
\end{array}
\qquad
\begin{array}{r}
0\ \ 1\ 0\ 1\ 0 \\
-\ 1\ \ 1\ 0\ 1\ 1 \\
\hline
\end{array}
\rightarrow
\begin{array}{r}
0\ \ 1\ 0\ 1\ 0 \\
+\ 0\ \ 0\ 1\ 0\ 0 \\
\hline
0\ \ 1\ 1\ 1\ 0 \\
\hline
1 \\
\hline
0\ \ 1\ 1\ 1\ 1
\end{array}
$$

The multiplication of two numbers in two's complement notation is a modification of the shift-and-add technique used for sign magnitude multiplication. Here, the multiplier forms an operation code, which tells whether to add, subtract, or do nothing for each step of the procedure. After each operation, a right shift occurs; then the next operation is performed.

Table 3-4 shows a typical two's complement multiplication. Two bit

Table 3-4.   Example of Two's Complement Multiplication.
Problem: 0 10101 × 0 101 = ?

Multiplier	Operation Code	Operation	Accumulative Result
0 10⌷1-⌷	10	Subtract 0 10101	0 00000 (Starting point = 0) + 1 01011 1 01011
0 1⌷01⌷	01	Shift right Add 0 10101	1 101011 + 0 10101 0 010101
0 ⌷10⌷1	10	Shift right Subtract 0 10101	0 0010101 + 1 01011 1 1000001
⌷0 1⌷01	01	Shift right Add 0 10101	1 11000001 + 0 10101 0 01101001 (Final product)

Result: 0 10101 × 0 101 = 0 1101001
+21 × +5 = +105

NOTE: A shift right is not required after the last operation. When shifting two's complement numbers to the right, the new bit shifted in is the same as the sign bit.

Operation Codes
00   Do nothing
01   Add
10   Subtract
11   Do nothing

groupings of the multiplier form the operation codes, which are interpreted as shown in the notes. After each operation, do a right shift, being careful to shift in the correct polarity bit as determined by the sign of the number. All addition and subtraction follows the previously established rules for two's complement arithmetic.

Comparison of this technique to the one shown in TABLE 3-2 reveals that the two methods are quite similar, except that two bit groupings are used to form the operation codes and that subtraction is one of the operations allowed for the two's complement case.

Division of two's complement numbers is also quite similar to the sign magnitude technique; the main difference is that the sign of the divisor and the sign of the result serve together to determine the operation code for each step. Also, addition and subtraction are both permitted operations for two's complement division, whereas the operations for sign magnitude division were limited to either subtract or do nothing (subtract 0).

An example of two's complement division is shown in TABLE 3-5. As

Table 3-5. Example of Two's Complement Division.
Problem: 0 1101001 ÷ 0 101 = ?

Operation	Accumulative Result	Sign Same As Divisor?	Quotient
	0 01101001 (Starting point)	—	
Shift left Subtract	0 11010010 +1 01100000 0 00110010	Yes	1 (MSB)
Shift left Subtract	0 01100100 +1 01100000 1 11000100	Yes	1
Shift left Add	1 10001000 +0 10100000 0 00101000	No	0
Shift left Subtract	0 01010000 +1 01100000 1 10110000	Yes	1
Shift left Add	1 01100000 +0 10100000 0 00000000	No	0 (LSB)
	Tentative Answer	1 .1010	
	Correction factor	1 00001	
		0 .10101	
	Shifted	0 10101	
	Result: 0 1101001 ÷ 0 101 = 0 10101		

NOTES. At start of problem, divisor must be greater than number being divided. If not, shift left until it is. At end of problem, shift answer same number of places.
Correction factor is always 1 000 ... 001 with last one being one place more than tentative answer.

Operation Codes

Sign Same As Divisor?	Operation	Quotient
Yes	Subtract	1
No	Add	0

with sign magnitude division, the starting point for the process is with the number being divided. Follow with a left shift, then the indicated operation. Repeat the procedure until a result of 0 is obtained or until the desired number of digits have been generated.

Of importance with the two's complement division technique is that the answer obtained is initially incorrect. To get the correct answer, 1 000 ... 001 must be added to the quotient. This correction can be thought of as simply changing the sign of the answer, then adding a one at the end. A comparison of TABLES 3-3 and 3-5 shows the similarities in the two techniques.

OPERATIONS WITH BCD CODES

A number represented in a BCD code is organized into four bit groups called *decades,* with each decade representing one digit of a decimal number. Within each decade, the rules for addition and subtraction are identical to the basic rules shown in TABLE 3-1. In standard binary notation, each adjacent binary bit represents an increasing power of two; carries or borrows are normal binary functions as previously described. In BCD, however, each decade uses only 10 of the possible 16 states available. As a result, special correction factors must be added or subtracted when using BCD to account for the unused states. Further, carries and borrows external to each decade must be based on the requirements for a decimal carry or borrow instead of a binary carry or borrow.

Table 3-6.   Corrections for Addition Using 8421 Code.

Decimal Sum	Uncorrected Carry	Uncorrected Sum	Corrected Carry	Corrected Sum	Correction
10	0	1010	1	0000	+6
11	0	1011	1	0001	+6
12	0	1100	1	0010	+6
13	0	1101	1	0011	+6
14	0	1110	1	0100	+6
15	0	1111	1	0110	+6
16	1	0000	1	0111	+6
17	1	0001	1	1000	+6
18	1	0010	1	1001	+6
19	1	0011	1	0101	+6

Addition with the 8421 Code

A common method of performing BCD addition is to add two numbers in binary fashion, then add or subtract an appropriate correction factor, if necessary. This is the method described here.

The initial addition will be correct using the 8421 code, provided the sum is not greater than 9. In cases where the sum does exceed 9, 6 must be added to convert the number to a corrected code. Thus, if the decimal sum is between 10 and 15, 6 must be added to the initial result and a carry generated for the next decade. If the decimal sum is greater than 15, a carry is generated automatically; however, it is still necessary to add 6 to the initial result. Table 3-6 indicates the corrections that must be made when performing addition with the 8421 code.

One disadvantage to correcting the 8421 code in the manner described is that each decade must be corrected, starting with the least significant decade, before the next decade can be corrected. The example here demonstrates addition using the 8421 code.

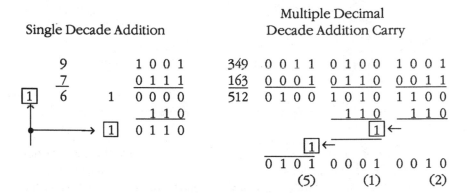

Single Decade Addition

Multiple Decimal Decade Addition Carry

Addition with the Excess-Three Code

When two excess-three numbers are added, the sum of the two excess 3s results in an excess 6. If the resultant decimal sum is 9 or less, a correction of -3 must be made to return to proper excess-three notation. However, if the decimal sum is greater than 9, the excess 6 will be larger than 15, thereby generating a carry from the decade and causing an overflow within the decade. The term *overflow* refers to the situation in which the sum requires more bits than are available within the decade. When this occurs, the sum shown by the four bits in the decade is incorrect and the carry bit must be sensed to properly correct the sum. Thus, in this case a correction of $+3$ is required to return excess-three notation. Table 3-7 shows the corrections to be made when performing addition with the excess-three code.

Table 3-7.   Corrections for Addition using Excess Three Code.

Decimal Sum	Uncorrected Carry	Uncorrected Sum	Corrected Carry	Corrected Sum	Correction
0	0	0110	0	0011	−3
1	0	0111	0	0100	−3
2	0	1000	0	0101	−3
3	0	1001	0	0110	−3
4	0	1010	0	0111	−3
5	0	1011	0	1000	−3
6	0	1100	0	1001	−3
7	0	1101	0	1010	−3
8	0	1110	0	1011	−3
9	0	1111	0	1100	−3
10	1	0000	1	0011	+3
11	1	0001	1	0100	+3
12	1	0010	1	0101	+3
13	1	0011	1	0110	+3
14	1	0100	1	0111	+3
15	1	0101	1	1000	+3
16	1	0110	1	1001	+3
17	1	0111	1	1010	+3
18	1	1000	1	1011	+3
19	1	1001	1	1100	+3

A convenient feature of the excess-three code is that the carry generated from the initial addition can be used to determine whether a +3 or −3 correction is to be made. If there is no carry, then −3 must be added. If a carry does occur, +3 is to be added. The following example demonstrates addition using the excess-three code:

	Single Decade Addition	Multiple Decade Addition	Decimal Carry

```
                                                                    Decimal
       Single Decade Addition      Multiple Decade Addition          Carry

       9            1 1 0 0     349  0 1 1 0   0 1 1 1    1 1 0 0
       7            1 0 1 0     163  0 1 0 0   1 0 0 1    0 1 1 0
 [1]   6      1     0 1 1 0     512  1 0 1 0   0 0 0 0    0 0 1 0
                 +  0 0 1 1        − 0 0 1 1 + 0 0 1 1  + 0 0 1 1
              [1]   1 0 0 1              1         1
                                    1 0 0 0   0 1 0 0    0 1 0 1
                                     (5)       (1)        (2)
```

Subtraction with the 8421 Code

The results for a subtraction using the 8421 code will be correct, providing the difference obtained is a positive number. However, if the difference is negative, a borrow must be performed from the next decade and a correction of -6 must be made to the binary difference to obtain the correct result. Table 3-8 lists the corrections to be made for subtraction using the 8421 code.

Table 3-8. Corrections for Subtraction Using 8421 Code.

Decimal Difference	Uncorrected Borrow	Uncorrected Difference	Corrected Borrow	Corrected Difference	Correction
-1	1	1111	1	1001	-6
-2	1	1110	1	1000	-6
-3	1	1101	1	0111	-6
-4	1	1100	1	0110	-6
-5	1	1011	1	0101	-6
-6	1	1010	1	0100	-6
-7	1	1001	1	0011	-6
-8	1	1000	1	0010	-6
-9	1	0111	1	0001	-6

As with 8421 addition, each decade must be corrected in order, starting with the least significant decade. An example of subtraction using 8421 code follows:

```
   45        0 1 0 0      0 1 0 1
 - 28        0 0 1 0      1 0 0 0
   17        0 0 1 0      1 1 0 1
                 - 1    - 0 1 1 0
             0 0 0 1      0 1 1 1
               (1)         (7)
```

Subtraction with the Excess-Three Code

An excess-three subtraction is similar to an 8421 subtraction except that different correction factors are required. If the difference is positive, the correction required is $+3$; if the difference is negative, then a correction of -3 is needed. Table 3-9 shows the corrections to be made for subtraction using the excess-three code.

**Table 3-9.　Corrections for Subtraction
Using Excess-Three Code.**

Decimal Difference	Uncorrected Borrow	Uncorrected Difference	Corrected Borrow	Corrected Difference	Correction
+9	0	1001	0	1100	+3
+8	0	1000	0	1011	+3
+7	0	0111	0	1010	+3
+6	0	0110	0	1001	+3
+5	0	0101	0	1000	+3
+4	0	0100	0	0111	+3
+3	0	0011	0	0110	+3
+2	0	0010	0	0101	+3
+1	0	0001	0	0100	+3
0	0	0000	0	0011	+3
−1	1	1111	1	1100	−3
−2	1	1110	1	1011	−3
−3	1	1101	1	1010	−3
−4	1	1100	1	1001	−3
−5	1	1011	1	1000	−3
−6	1	1010	1	0111	−3
−7	1	1001	1	0110	−3
−8	1	1000	1	0101	−3
−9	1	0111	1	0100	−3

Note that a borrow is generated if the difference is negative and there is no borrow when the difference is positive. An example of subtraction using excess-three code is shown here:

```
  45        0 1 1 1        1 0 0 0
- 28      - 0 1 0 1        1 0 1 1
  17        0 0 1 0        1 1 0 1
                 - 1
          + 0 0 1 1      - 0 0 1 1
            0 1 0 0        1 0 1 0
              (1)            (7)
```

Summary

Binary arithmetic procedures vary depending on the type of notation being used. For pure binary notation, all the operations are comparable to decimal operations. Operations with sign magnitude notation numbers are done the same way as for pure binary numbers, except that care must be taken to observe the sign and to place the correct sign in front of the answer. One's complement addition is done the same as for pure binary numbers, except that an end-around carry is used. One's complement subtraction requires inverting and then adding. Two's complement addition avoids the end-around carry; any carry bit is ignored. Two's complement numbers are subtracted by complementing and then adding. Multiplication and division using two's complement notation involve operation codes which tell whether to add or subtract at any given time. Operations with BCD codes are normally done one decimal digit at a time, with carries and corrections made on individual digit groups.

Questions

1. Explain the method for multiplying binary numbers when one of the numbers is an even power of 2.

2. Assume that decimal 9 is to be subtracted from decimal 15. Explain how this is done using **a.** pure binary notation **b.** sign magnitude notation **c.** one's complement notation **d.** two's complement notation **e.** 8421 notation **f.** excess-three notation

3. Assume that decimal -3 is to be subtracted from decimal 12. Explain how this is done using **a.** pure binary notation **b.** sign magnitude notation **c.** one's complement notation **d.** two's complement notation **e.** 8421 notation **f.** excess-three notation

4. Explain why binary multiplication is sometimes called *shift and add.*

5. Explain why binary division is sometimes called *shift and subtract.*

Problems

1. Given $A = 11001$ and $B = 101$, calculate **a.** $A + B$ **b.** $A - B$ **c.** AB **d.** A/B

2. Repeat Problem 1 for $A = 11110$ and $B = 110$.

3. In sign magnitude notation, $C = 0\ 11011$ and $D = 1\ 1001$. Calculate
    a. $C + D$ **b.** $C - D$ **c.** CD **d.** C/D

4. Repeat Problem 3 for $C = 1\ 11000$ and $D = 0\ 110$.

5. In one's complement notation, $E = 0\ 1001$ and $F = 1\ 0011$. Find
    a. $E + F$ **b.** $E - F$

6. Repeat Problem 5 for $E = 0\ 1111$ and $F = 1\ 1001$.

7. Repeat Problem 5, assuming that E and F are in two's complement form.

8. Repeat Problem 6, assuming that E and F are in two's complement form.

9. In two's complement notation, $G = 0\ 100011$ and $H = 0\ 111$. Find
    a. GH **b.** G/H

10. Repeat Problem 9, with $G = 0\ 111100$ and $H = 0\ 1100$.

11. In 8421 notation, $J = 0101\ 1001\ 0111$, $K = 0010\ 0001\ 0110$. Find $J + K$.

12. Repeat Problem 11 with $J = 1001\ 0011\ 0111$ and $K = 0010\ 1001\ 0101$.

13. Using the same values as in Problem 11, **a.** convert J and K to excess-three form **b.** perform the addition **c.** correct the addition

14. Using the same values as in Problem 12, **a.** convert J and K to excess-three form **b.** perform the addition **c.** correct the addition

15. Using the same values as in Problem 11, **a.** convert J and K to excess-three form **b.** calculate $J - K$ **c.** correct the subtraction.

16. Using the same values as in Problem 12, **a.** convert J and K to excess-three form **b.** calculate $J - K$ **c.** correct the subtraction

LOGIC GATES

AFTER YOU COMPLETE this chapter, you will be able to:

☐ Understand positive logic
☐ Follow the procedures used in Boolean algebra
☐ Understand the meaning of different logic functions
☐ Understand logic symbols, truth tables, and schematics for logic functions
☐ Distinguish between discrete components and integrated circuits

*D*igital circuits perform the binary arithmetic operations described in chapters 1, 2, and 3 with the binary digits 1 and 0. Similarly, there are other types of functions that can be performed with digital circuits using only these two binary numerals. These functions are referred to as *logic* functions. Logic functions can all be described in terms of algebraic statements. The terms in these algebraic expressions all have the designation of either 1 or 0. Other expressions for 1 or 0 are "on" or "off," and "true" or "false." (See TABLE 4-1.)

Table 4-1. Relationships of Binary States: Positive Logic.

1	True	On	High	+ Volts
0	False	Off	Low	0 Volts
				(or − Volts)

For convenience in relating circuit operation to binary 1s and 0s, the *logical 1* is assumed to be the most positive voltage and the *logical 0* represents the most negative voltage. This relationship is known as *positive logic* and is used as a convention throughout this book. The reverse situation (1 designates a low and 0 designates a high) is *negative logic*.

BOOLEAN ALGEBRA

The algebra used to symbolically describe logic functions is *Boolean algebra*. As with ordinary algebra, the letters of the alphabet can be used to represent variables. The primary difference is that Boolean algebra variables can have only the values 1 or 0.

There are three connecting symbols used in Boolean algebra: the *equals sign* (=), the *plus sign* (+), and the *multiply symbol* (•). They are defined as follows:

The equals sign refers to a standard mathematical equality. That is, the logical value on one side of the sign is identical to the logical value on the other side of the sign. Hence, given that there are two logical variables A and B such that $A = B$, it follows that if $A = 1$ then $B = 1$, and if $A = 0$ then $B = 0$.

The plus sign in Boolean algebra refers to the logical OR function. If $A + B = 1$, the meaning is that either A or B represents the logical value 1. Therefore, when the statement $A + B = 1$ appears, then either $A = 1$ or $B = 1$ or both equal 1.

The logical multiply sign is called the AND function. If it is given that $A \cdot B = 1$, the interpretation is that both A and B represent the logical value 1. Thus, an algebraic equation of the form $A \cdot B = 1$ means that $A = 1$ *and* $B = 1$. If either A or B is not a logical 1, then $A \cdot B$ cannot possibly be a logical 1. The function $A \cdot B$ is often written AB, omitting the dot for convenience.

Boolean algebra also uses a logical NOT operation, indicated by a bar over the variable. The NOT has the effect of inverting (complementing) the logical value. Thus, if $A = 1$, then $\overline{A} = 0$. Additionally, parentheses are frequently used, as in regular algebra, to indicate groupings of logical operations.

Used in conjunction with Boolean algebra are truth tables and logic symbols. A *truth table* simply shows all of the possible values for the inputs to a function, then shows the resultant output for each combination of inputs. The purpose of truth tables is not to determine whether or not circuits are "lying." Rather, they verify "truth" in the logical sense—that a given set of input conditions will produce a "true" output, while some other set of input conditions produces a "false" output. (A true state is indicated by the digit 1, and a false state is indicated by the digit 0.) A *logic symbol* is a graphic way of indicating a particular logic function. Logic symbols are explained in more detail shortly.

LOGIC FUNCTIONS

Because Boolean algebra is used on elements having two possible stable states, it becomes very useful in analyzing logic (switching) circuits. The reason is that a switching circuit can be in only one of two possible states; that is, it is either a closed circuit or an open circuit—it's either turned on or it's turned off. As stated earlier, these two states are represented by the binary values of 1 and 0.

The logic functions explained in this section are:

- Equality
- NOT (Inverter)
- OR
- NOR
- AND
- NAND
- EX-OR
- EX-NOR

Logical Equality (*A* = *B*)

The logic function called *equality* actually means that two points or two signals are functionally identical. As seen in FIG. 4-1, a piece of wire or some arbitrary buffer amplifier is a typical condition of equality in a digital circuit. (A *buffer* is a circuit or device that does not alter the signal going through it and provides isolation or some other function.) The logic con-

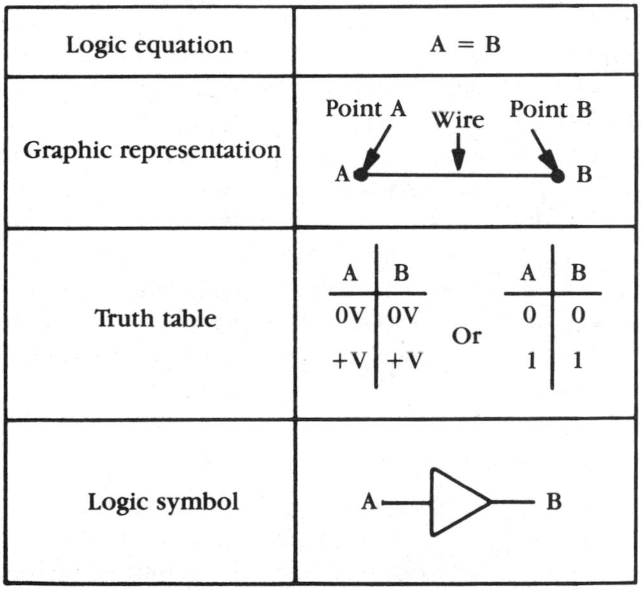

Logic equation	$A = B$
Graphic representation	Point A Wire Point B A●————————————●B
Truth table	A | B A | B 0V | 0V Or 0 | 0 +V | +V 1 | 1
Logic symbol	A ——▷—— B

4-1   Function: logical equality. Symbol denotes a noninverting amplifier or buffer (performs no logic function).

dition at point *A* is identical to the logic condition at point *B*. Also, the truth table shows the relationship of input to output conditions. Namely, if *A* = 0 then *B* = 0, and if *A* = 1 then *B* = 1. (Recall the preceding paragraph defining the basic equality function for Boolean algebra.)

Inverter or NOT Function ($\overline{A} = B$)

The inverter is probably the simplest and most basic digital circuit. Yet inverters occur as a part of almost all other more complicated circuits. The inverter's function in a digital circuit is very simple. It performs the logical NOT operation.

The logical NOT operation can be thought of, from a circuit standpoint, as an inverter. Whatever the input to the inverter, the output assumes the opposite polarity. If the input to an inverter is a logic 1, its output is represented by a logic 0. Similarly, if the inverter input is a logic 0, then a 1 is required as an output.

Normally, an inverter designed for use in logic circuits is a saturated-mode transistor switch. That is, the transistor acts very much like a mechanical switch. In the OFF condition, the transistor is cut off, and current flow from emitter to collector is very small. Hence, the switch is effectively open. In the ON condition, the transistor is driven well into saturation and the transistor is essentially a short circuit between the emitter and collector terminals. Here, the "switch" is considered closed. A typical saturated-mode inverter circuit is shown in FIG. 4-2.

Initially, assume that a ground (0V) is applied at input A, representing a logic 0. Resistors R_B and R_K form a voltage divider between ground and the negative voltage, thus holding the base of npn transistor Q1 negative. This negative potential causes Q1 to be cut off, and the output at point B is essentially + V. This represents the logic 1 state and is the correct operation for a NOT function as shown in the truth table.

If, on the other hand, a sufficiently large positive voltage is applied at point A, the base of transistor Q1 will become positive, causing the transistor to conduct heavily. The output at point B will therefore approach ground, the logic 0 level. Again, this action corresponds to the truth table for a logical NOT function.

OR Function ($A + B = F$)

An OR gate is the name given to a digital switching circuit that performs the logical OR function. A very simple form of an OR gate is several diodes connected such that the diodes are normally biased off, representing the logic 0 condition. When a logic 1 is applied at either diode, that diode is forward biased, thus producing a logic 1 output. This circuit, consisting of

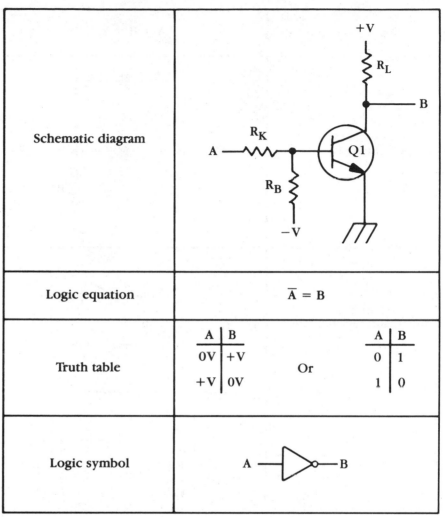

Schematic diagram	
Logic equation	$\overline{A} = B$
Truth table	(see table below)
Logic symbol	A ▷o— B

Truth table:

A	B
0V	+V
+V	0V

Or

A	B
0	1
1	0

4-2 Function: inverter or NOT. The circle at the output of the symbol indicates inversion.

two diodes and a resistor connected to $-V$, is shown in FIG. 4-3. If point A or point B is grounded, the associated diode (CR1 or CR2) is forward biased and a ground is present at point F. Since ground is more positive than $-V$, the ground represents a logic 1 and $-V$ represents the logical 0. The only time that output F is at $-V$ (logical 0) is when both A and B inputs are sufficiently negative to reverse-bias both diodes. This action corresponds to the truth table defined for an OR function.

To clarify the meaning of an OR function, consider the simplified

Logic equation	$A + B = F$
Schematic diagram	
Simplified schematic representation	
Truth table	
Logic symbol	

A	B	F		A	B	F
0V	0V	0V		0	0	0
0V	+V	+V	Or	0	1	1
+V	0V	+V		1	0	1
+V	+V	+V		1	1	1

4-3   Function: OR.

schematic representation in FIG. 4-3. Both switches are connected to some positive voltage, + V. The positive voltage will be present at point F if switch A is closed or if switch B is closed. If neither switch is closed, the positive voltage will not be present at point F. The truth table tells the same story in terms of 1's and 0's. For the A and B columns, 1 means a closed switch and 0 means an open switch. In the F column, 1 refers to the positive voltage being present and 0 refers to the absence of the positive voltage. A comparison of the truth table to the statements given in the OR function definition shows that the two are just different ways of indicating the same thing.

The OR function can be stated as follows: If *A* or *B* (or both) is high, the output is high.

NOR Function ($\overline{A + B} = F$)

The NOR function is a combination of the OR and the NOT functions. NOR comes from the expression "*not or*." The outputs of a NOR function are simply the inversion of those produced by an OR circuit, as the truth table in FIG. 4-4 shows. Both inputs must be low to obtain a high; all other com-

Logic equation	$\overline{A + B} = F$		
Truth table	(see table below)		
Logic symbol	(see figure below)		

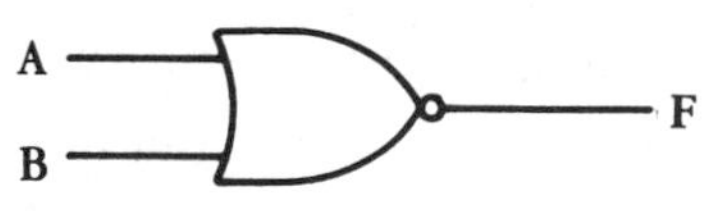

A	B	F		A	B	F
0V	0V	+V		0	0	1
0V	+V	0V	Or	0	1	0
+V	0V	0V		1	0	0
+V	+V	0V		1	1	0

4-4　Function: NOR.

binations produce a low. Note the symbol—it's simply an OR symbol with the characteristic circle at the output, indicating inversion.

The NOR function can be stated thus: if neither *A* nor *B* is high, the output is high.

AND Function (*A* · *B* = *F*)

The AND gate performs the logical AND function. This gate can be formed in very much the same manner as the OR gate, but with the diodes reversed and with the resistor returned to + V. An AND gate circuit of this type is shown in FIG. 4-5. The voltage that represents a logic 1 in this circuit is + V, while 0V represents a logic 0. If a ground is present at either point A or point B, one of the diodes will be forward biased, causing point F to also be at ground potential (logic 0). It is required that both points A and B be positive to reverse-bias diodes CR1 and CR2. In this condition only, output *F* will be at + V. The truth table shows that the resultant circuit operation is that required for an AND function.

The two switches used to represent the OR function were wired in parallel. If instead two switches are wired in series, as shown in FIG. 4-5, a simplified representation of an AND function is obtained. Both switches *A* and *B* must be closed to obtain a positive voltage at point F. If either switch is open, the positive voltage will not be present at the output. The truth table reflects the definitions previously stated.

The AND function can be stated thus: If *A* and *B* are high, the output is high.

NAND Function ($\overline{AB}$ = *F*)

As with the NOR function, the NAND circuit inverts the expected outputs of the AND function. Expectedly, NAND is short for "*not and.*" FIGURE 4-6 shows the truth table and symbol for the NAND function.

The NAND function can be stated thus: If *A* and *B* are high, the output is low.

Exclusive-OR Function

The exclusive-OR function (EX-OR) is another function of interest because it is commonly used in a number of applications. The EX-OR merely means the OR of either *A* or of *B* but exclusive of each other, thus not including the case of both *A* and *B* being true at the same time. (See FIG. 4-7.) EX-OR is represented in equations by the symbol ⊕ and is arithmetically equal to modulo-two addition. Recall from chapter 2 that modulo-two sum is simply a binary sum with no carry. The EX-OR circuit is quite useful where arithmetic operations are to be performed. Also, an EX-OR

Logic equation	AB = F or A• B = F
Schematic diagram	
Simplified schematic representation	
Truth table	
Logic symbol	

4-5 Function: AND.

Logic equation	$\overline{AB} = F$	
Truth table	 A B | F 0V 0V | +V 0V +V | +V +V 0V | +V +V +V | 0V Or	A B | F 0 0 | 1 0 1 | 1 1 0 | 1 1 1 | 0
Logic symbol		

4-6. Function: NAND.

Logic equation	$A \oplus B = F$	
Truth table	 A B | F 0V 0V | 0V 0V +V | +V +V 0V | +V +V +V | 0V Or	A B | F 0 0 | 0 0 1 | 1 1 0 | 1 1 1 | 0
Logic symbol		

4-7. Function: EX-OR.

function represents a simple parity circuit. That is, the output of the EX-OR is a 1 if an odd number of bits are 1 and is a 0 if an even number of bits are 1. This is identical to the even parity scheme described in chapter 2. Odd parity is the inverse of the EX-OR function, known as the EX-NOR function.

The EX-OR function can be stated thus: If exclusively *A* or exclusively *B* is high, the output is high. In other words, if *A* and *B* are different, the output is high.

EXCLUSIVE-NOR Function

Exclusive-NOR (EX-NOR) logic produces the opposite outputs of the EX-OR function (see FIG. 4-8 for the truth table and symbol). As stated before, it can be used as an odd parity function.

Logic equation	$\overline{A \oplus B} = F$
Truth table	<table><tr><td>A</td><td>B</td><td>F</td><td></td><td>A</td><td>B</td><td>F</td></tr><tr><td>0V</td><td>0V</td><td>+V</td><td rowspan="4">Or</td><td>0</td><td>0</td><td>1</td></tr><tr><td>0V</td><td>+V</td><td>0V</td><td>0</td><td>1</td><td>0</td></tr><tr><td>+V</td><td>0V</td><td>0V</td><td>1</td><td>0</td><td>0</td></tr><tr><td>+V</td><td>+V</td><td>+V</td><td>1</td><td>1</td><td>1</td></tr></table>
Logic symbol	

4-8 Function: EX-NOR.

The EX-NOR function can be stated thus: If exclusively *A* nor exclusively *B* is high, the output is high. Or, if *A* and *B* are the same, the output is high.

CHARACTERISTICS OF GATES

Logic gates have certain properties that engineers must take into account when designing digital circuitry. Considerations include requirements in speed, efficiency, power, and loading. This section briefly describes the gate characteristics of propagation delay, power dissipation, noise immunity, and loading considerations. These are some of the many qualities that are specified in manufacturers' data sheets.

Gate Propagation Delay   Propagation delay is a very important characteristic of logic circuits because it limits the speed (frequency) at which the circuits can operate. The terms *low speed* and *high speed,* when applied to logic circuits, refer to the propagation delays; the shorter the propagation delay, the higher the speed of the circuit.

The maximum frequency of input pulses at which the gate can operate is inversely related to the propagation delay. The greater the delay, the lower the maximum frequency at which the gate will function and vice versa.

Noise Immunity   Every logic circuit has certain limits on the values of the voltages it will operate properly within. The *dc noise margin* of a logic gate is a measure of its *noise immunity,* which is a gate's ability to withstand fluctuations of the voltage level (noise) with which it must operate. Common sources of noise are variations of the dc supply voltage, ground noise, magnetically coupled voltages from adjacent lines, and radiated signals. The term *dc noise margin* applies to noise voltages of relatively long duration compared to a gate's response time.

Loading Considerations   Digital systems typically have many types of digital ICs interconnected to perform various functions. In these situations, the output of a logic gate might be connected to the inputs of several other gates, so the *load* on the driving gate becomes an important factor.

The *fan-out* of a gate is the maximum number of inputs of the same IC family (see chapter 5) that the gate can drive while maintaining its output levels within specified limits. That is, the fan-out specifies the maximum loading that a given gate is capable of handling.

LOGIC GATE APPLICATIONS

This section shows two simple examples that demonstrate how logic gates might be applied to practical situations.

AND Gate Application

In a simple application, and AND gate can be used to detect the existence of a specified number of conditions and in response activate an appropriate action.

For example, an automobile's safety system might require that an audible signal be produced to warn the driver a seat belt is not engaged. The conditions are that the ignition switch be on, the seat belt be unbuckled, and the warning signal be on for a specified period and turn off automatically. The first two conditions can be represented by switch positions and the third by a timer circuit.

Figure 4-9 shows an AND gate whose high output activates a buzzer when these three conditions are met on its inputs. When the ignition switch (represented by S_1 in the figure) is on, a high is connected to the gate input A. When the gate is not properly buckled, switch S_2 is off and a high is connected to the gate input B. At the instant the ignition switch is turned on, the timer is activated and produces a high on gate input C. The resulting high gate output activates the alarm. After a specified time, the timer circuit's input goes low, disabling the AND gate and turning off the alarm.

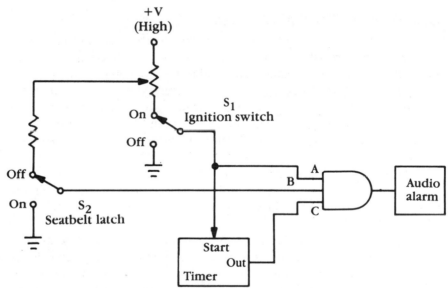

4-9   Example of an AND gate application.

OR Gate Application

As an example application of an OR gate, assume that in a room with three doors, an indicator lamp must turn on when any of the doors is open.

The sensors are switches that are open when a door is ajar or open. This open switch creates the high level for the OR gate input as shown in FIG. 4-10. If any or all of the doors are open, the gate output is high. This high is then used to light the indicator lamp.

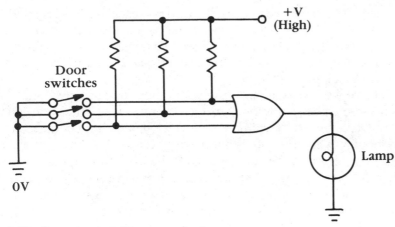

4-10   Example of an OR gate application.

ABOUT INTEGRATED CIRCUITS

All of the circuits described thus far were shown schematically as consisting of individual resistors, transistors, and diodes. These components are referred to as *discrete components*. Another type of component is the integrated circuit (IC). Using modern manufacturing techniques, an entire group of active circuit components can be manufactured on a single chip or semiconductor substrate. Each component could not be removed and looked at separately, but rather the entire IC assembly can be considered as one new component.

The obvious advantages over discrete components are that integrated circuits are much smaller than their discrete counterparts, they are almost as inexpensive to mass-produce as discretes, and there are fewer parts that can fail.

Basically, integrated circuits are constructed much the same as individual diodes and transistors. In A of FIG. 4-11, that each integrated circuit typically starts out as a single crystal chip of silicon, either positively (p) or negatively (n) doped. This crystal chip is referred to as the *bulk,* or *substrate,* material. Into the substrate, additional *p* and *n* dopants are added regionally through a diffusion process to form the desired components.

Sketch B shows how an IC resistor is formed from just a single n-type region added to the p-type substrate. Metal contacts are jointed to two physically separate portions of the n-type material, using the resistance of the material to form the actual resistor. The resistor obtained is dependent primarily on the geometry of the n-type region. The actual resistivity of the material is usually predetermined; varying resistances are obtained by changing the length and width of the region, as required. From the given

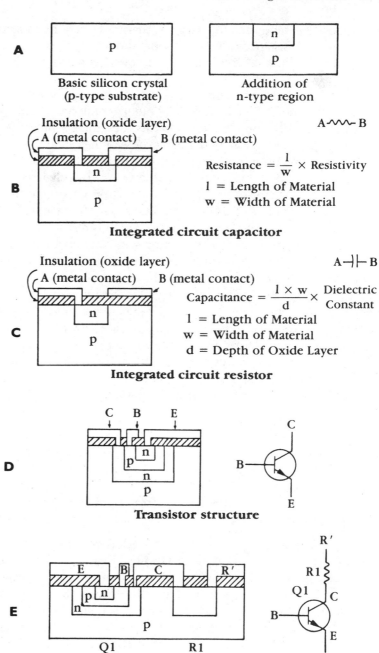

A

Basic silicon crystal
(p-type substrate)

Addition of
n-type region

B

Insulation (oxide layer)

A (metal contact)   B (metal contact)

A $-\!\!\wedge\!\!\wedge\!\!\wedge\!\!-$ B

$$\text{Resistance} = \frac{l}{w} \times \text{Resistivity}$$

l = Length of Material
w = Width of Material

Integrated circuit capacitor

C

Insulation (oxide layer)

A (metal contact)   B (metal contact)

A $-\!|\!\vdash$ B

$$\text{Capacitance} = \frac{l \times w}{d} \times \begin{array}{l}\text{Dielectric}\\\text{Constant}\end{array}$$

l = Length of Material
w = Width of Material
d = Depth of Oxide Layer

Integrated circuit resistor

D

C   B   E

Transistor structure

E

E   B   C   R′

Q1   R1

Complete integrated circuit

4-11   Integrated circuit structure.

equation, it can be seen that large resistances tend to be long and narrow, while small resistances tend to be short and wide. An oxide layer is placed between the body of the chip and the metal contacts for insulation and circuit isolation.

To form an IC capacitor, sketch C of the figure shows how the oxide layer forms the dielectric material. Note that only one metal contact actually touches the n-type material. In this case, the geometry of the n-type material, in conjunction with the thickness of the oxide layer, is the determining factor in the capacitance obtained. From the equation, it is apparent that since the depth of the oxide layer is relatively fixed, the physical size of the capacitor grows rapidly in direct proportion to the amount of capacitance required. This is an important fact to keep in mind. Inherently, IC capacitors require fairly large areas and thus reduce the amount of circuitry that can be fabricated on a given semiconductor chip.

A simple transistor structure is shown in sketch D to illustrate the similarity in basic structure of the integrated circuit to the transistor. Finally, in sketch E, a composite integrated circuit made up of a transistor and a resistor is shown. Physically, the structure is much the same as if the chips from sketches B and D were side by side. However, as was stated, the integrated circuit is formed on only one chip, thereby reducing the number of components, in this case from two to one. Very complex structures are made up in the same manner. Modern integrated circuits can contain several thousands of transistors and resistors on one chip.

Summary

In digital circuits, only two states are used. These may be high or low, on or off, true or false, or 1 or 0. In a positive logic system, a high signal has a positive voltage, and a low signal has zero or a negative voltage. Boolean algebra is the mathematics of logic. The four primary symbols are = (meaning equality), + (meaning OR), •, the multiplication sign (meaning AND), and ⁻ (meaning NOT). Combinations of these symbols create functions called NOR, NAND, EX—OR, and EX—NOR. Logic gates exhibit such characteristics as gate propagation delay, noise immunity, and loading. All of these affect gate operations. Resistors, diodes, and transistors are examples of discrete components. Integrated circuits contain many of these individual items constructed as a single unit.

Questions

1. What is the difference between positive logic and negative logic?

2. Write a logic equation for each of the following:
   a. $A = \text{NOT } B$ **b.** $A \text{ OR } B = F$ **c.** $A \text{ AND } B = F$

3. What is meant by *gate propagation delay?* Why is it important?

4. What is meant by *noise immunity?*

5. What does the term *loading* refer to?

6. What is the difference between discrete components and integrated circuits?

Problems

1. Draw a truth table for each of the following: **a.** $A = B$
   b. $A + B = F$ **c.** $AB = F$ **d.** $A \text{ EX-OR } B = F$

2. Draw a truth table for each of the following: **a.** $A = \text{NOT } B$
   b. $A \text{ NOR } B = F$ **c.** $A \text{ NAND } B = F$ **d.** $A \text{ EX-NOR } B = F.$

3. Draw logic symbols for each of the expressions in Problem 1.

4. Draw logic symbols for each of the expressions in Problem 2.

5 ◆ LOGIC FAMILIES

AFTER YOU COMPLETE this chapter, you will be able to:

☐ Describe the different logic families and explain the advantages and disadvantages of each
☐ Understand what is meant by a high-impedance output state
☐ Understand the difference between depletion mode and enhancement mode MOSFETs

Before we delve into the circuits these basic gates make up, this chapter explains some of the types of logic used to create such circuits. The various approaches to digital logic design are called *logic families*.

Many different types of logic families exist and, depending on the application, circuits in a particular piece of equipment may be selected from one or more of these families. Factors influencing selection of a particular logic family include speed of operation, noise immunity, versatility, power consumption, size, and cost, among other considerations. This chapter introduces the commonly used logic families and indicates the distinguishing features of each.

RELAY LOGIC

Probably the simplest logic family, in terms of electronic components, is relay logic. Relay logic uses the relay windings as logic inputs and considers the resultant open or closed relay contacts as the logic 1 and logic 0 outputs.

The inverter circuit shown in FIG. 5-1 uses an input A at the relay winding to control whether or not the circuit through R_L is open or closed. Initially, input A is low (at ground potential) and the relay is in its normally closed position. In this condition, the load is provided with a ground and current flows through R_L. When logic signal A goes high ($+$ V is applied), the relay opens and so does the path for current flow through R_L. The condition where current flows through the load is considered the logic 1 state; absence of current flow denotes the logic 0 state. From the defini-

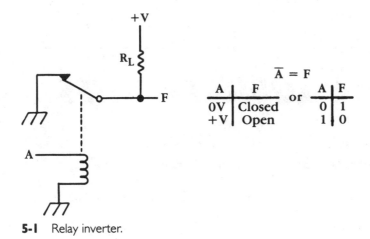

5-1   Relay inverter.

tions it can be seen that the circuit of FIG. 5-1 performs the inversion (NOT) function.

It is important when dealing with relay logic to determine whether a normally open or normally closed relay is used. In the diagram, the winding is placed in the direction that the contact will move, thus indicating that when the relay is energized, the contacts will open. Hence, this example shows a normally closed relay. Resistance R_L in the example typically includes the resistance of other relay windings that are being driven by the logic function. The value of R_L is selected to be large enough to prevent excessive current draw from the power supply but small enough to ensure that sufficient current is available to energize relays in the load circuit.

Examples of relay AND and relay OR circuits are shown in FIG. 5-2. In each case, whenever a closed circuit to ground is obtained, current can flow through R_L and the output is considered to be in the logic 1 state. Examination of the truth tables for each of the circuits shows that the circuits do perform the AND and OR functions as defined in chapter 4.

A more complex example of relay logic, using three input variables, is shown in FIG. 5-3. Here, the function $(A + B)C = F$ is shown with its associated truth tables. Multivariable truth tables are formed by tabulating all possible combinations of the input variables and writing down the resultant output for each combination.

Relay logic has low power dissipation when the relay is not energized and does not require a regulated power supply. This type of logic is ideal in a high-noise environment, such as a manufacturing facility because it is virtually immune to noise transients. Also, relays are capable of switching large currents required in some industrial applications. Several disadvantages to relay logic are: its large size relative to transistors and integrated

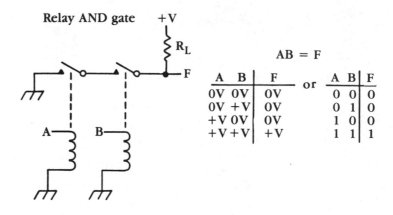

Relay AND gate   +V

$$AB = F$$

A B	F		A	B	F
0V 0V	0V	or	0	0	0
0V +V	0V		0	1	0
+V 0V	0V		1	0	0
+V +V	+V		1	1	1

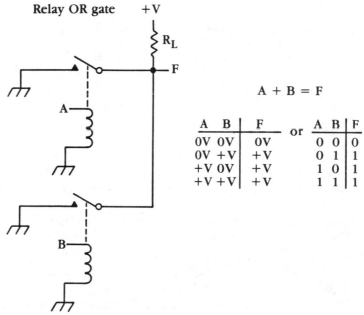

Relay OR gate   +V

$$A + B = F$$

A B	F		A	B	F
0V 0V	0V	or	0	0	0
0V +V	+V		0	1	1
+V 0V	+V		1	0	1
+V +V	+V		1	1	1

5-2   Examples of relay AND and OR gates.

circuits, its relatively slow switching speeds, and the adverse effects of contact bounce when the relay first opens or closes.

DIODE LOGIC

Diode AND and OR gates were described in the section on basic logic functions in chapter 4. Therefore, only a brief review of diode logic is included here.

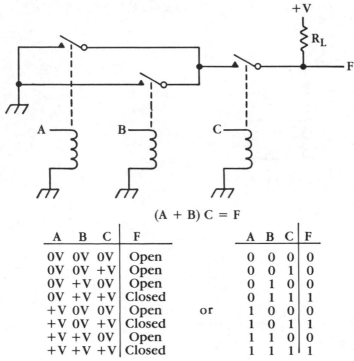

$$(A + B)\ C = F$$

A	B	C	F
0V	0V	0V	Open
0V	0V	+V	Open
0V	+V	0V	Open
0V	+V	+V	Closed
+V	0V	0V	Open
+V	0V	+V	Closed
+V	+V	0V	Open
+V	+V	+V	Closed

or

A	B	C	F
0	0	0	0
0	0	1	0
0	1	0	0
0	1	1	1
1	0	0	0
1	0	1	1
1	1	0	0
1	1	1	1

5-3   Complex relay logic function.

A diode provides a very low impedance when biased in the forward direction and a very high impedance when biased in the reverse direction. This two-level action is similar to the opening and closing of relay contacts just described.

However, compared to relay circuits, diodes can operate at lower voltages and have lower overall power dissipation. Furthermore, diodes have very fast switching times and are physically quite small.

The main disadvantages to diode logic are: the impedance of a diode circuit is quite sensitive to loading, thus providing very poor drive capability, and each series diode required produces a voltage drop and a resultant shift in the typically low-output voltage levels. Thus, diode logic does not lend itself well to series logic functions, and transistor power amplifiers must be used frequently to provide drive power and restoration of the voltage level.

A multilevel diode logic circuit is shown in FIG. 5-4. In this figure, there are two levels of series diode drop. Assuming silicon diodes are used, each diode drops approximately 700 millivolts (0.7V), making a total

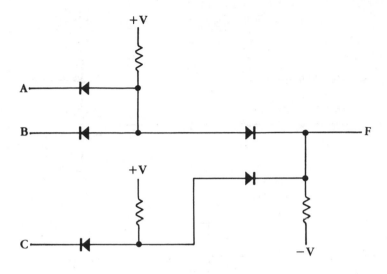

$$(A\ B) + C = F$$

A	B	C	F		A	B	C	F
−V	−V	−V	−V		0	0	0	0
−V	−V	+V	+V		0	0	1	1
−V	+V	−V	−V		0	1	0	0
−V	+V	+V	+V		0	1	1	1
+V	−V	−V	−V	**or**	1	0	0	0
+V	−V	+V	+V		1	0	1	1
+V	+V	−V	+V		1	1	0	1
+V	+V	+V	+V		1	1	1	1

5-4 Multilevel diode logic.

drop in the circuit of 1.4V. Therefore, it does not take many levels of series diode logic to produce a significant voltage drop.

DIODE-TRANSISTOR LOGIC

A basic form of diode-transistor logic (DTL) utilizes a diode AND gate followed by a transistor inverter to form a NAND gate, as shown in FIG. 5-5. A DTL NOR gate is formed in a similar manner, by adding a transistor inverter to a diode OR gate, as shown in FIG. 5-6. This logic family provides moderately high speed operation and good fanout (drive) capability. Fabrication of DTL gates can be from individual transistors and diodes or in the form of integrated circuits. One of the main disadvantages to the DTL circuits shown is that both configurations require the use of both positive and negative voltage power supplies for operation.

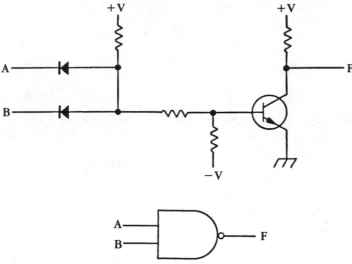

5-5   Basic DTL NAND gate configuration.

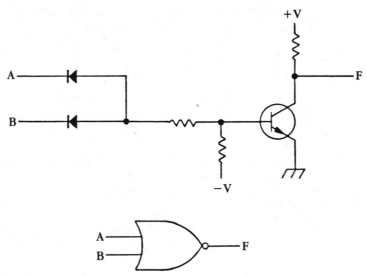

5-6   Basic DTL NOR gate configuration.

A modified version of a DTL NAND gate, which is more suitable for IC applications, is shown in FIG. 5-7. Here, resistors R1 and R2 form a bias feedback network that provides additional stability and temperature range. Furthermore, with the addition of a second transistor, increased

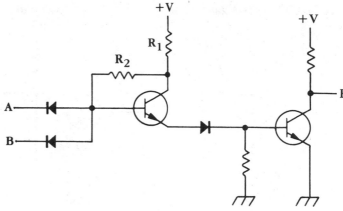

5-7 Modified DTL NAND gate.

drive capability is obtained and the requirement for a negative-voltage power supply is eliminated.

The DTL family is probably the easiest logic to use, and is also the cheapest form of logic available for medium-speed applications.

DIRECT-COUPLED TRANSISTOR LOGIC

Logic gates can be made by direct interconnection of transistors. This type of logic is called *direct-coupled transistor logic* (DCTL) and was one of the first types of logic made into integrated circuits. Typical NAND and NOR gates using DCTL circuits are shown in FIG. 5-8. The DCTL gate is simple to make, needing few parts, and is easy to produce in IC form. Also, DCTL requires only a single low-voltage power supply.

Typically, DCTL circuits provide a voltage swing on the order of several volts when operating from a + 3V power supply. Because of the lack of a turnoff bias for the transistor and because of the small voltage swings, the noise immunity of DCTL circuits is poor. In their off states, the transistors in DCTL circuits operate very near the edge of conduction. For this reason, very good grounding is required; otherwise, locally generated noise is apt to cause spurious outputs, which can trigger subsequent logic stages.

Another problem with DCTL is *current hogging,* which results when the bases of two or more transistors are driven directly from the collector of a single driver stage. If one transistor happens to turn on earlier than the others, the output transistor might be clamped to a value that is insufficient to turn on the other transistors. Because of this current-hogging characteristic, the drive capability of DCTL is quite limited. Also, in order

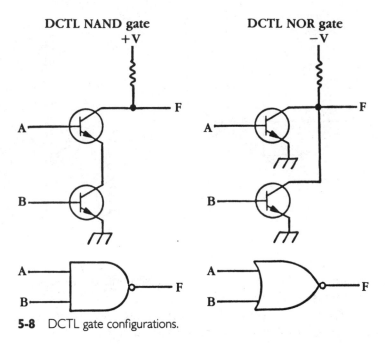

5-8   DCTL gate configurations.

to minimize current hogging, transistors must be carefully selected to have very nearly identical turn-on voltages.

RESISTOR-TRANSISTOR LOGIC

Resistor-transistor logic (RTL) utilizes only resistors and transistors to make up logic gates. RTL gates can be configured several different ways, as shown in FIG. 5-9. One configuration is similar to DTL in that the resistors perform the logic functions and the transistors serve as inverters. Another configuration is similar to DCTL, but with base resistors added to eliminate current hogging.

Because of the few components required, RTL circuits are simple and reliable. Also, these circuits have relatively low power consumption and are inexpensive. Disadvantages of the RTL family are: their slow switching speeds because of their tendency to be driven deeply into saturation, and their poor drive capability compared to other logic families.

RESISTOR-CAPACITOR-TRANSISTOR LOGIC

The RTL transistor switching circuits just described suffer from slow switching speeds because of their tendency to draw excessive base current. This base current drives the transistor into the saturation region, resulting in an accumulation of stored charge. The time to remove the

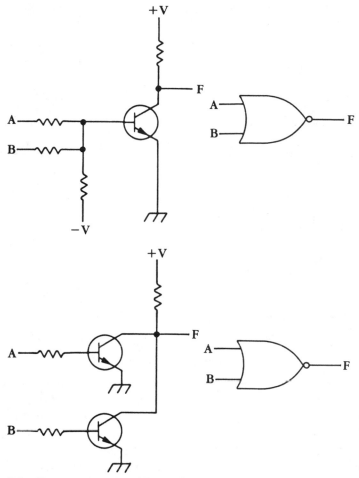

5-9 Two versions of an RTL NOR gate.

stored charge as the transistor attempts to change states is called the *storage delay time*. One method of reducing storage delay time is to add up a speedup capacitor in parallel with the base resistor, as shown in FIG. 5-10. The speedup capacitor stores the charge instead of the transistor, thus allowing the transistor to switch out of saturation faster. When a resistor and capacitor are used in this manner, the resultant logic circuit is called *resistor-capacitor-transistor logic* (RCTL).

One disadvantage to RCTL is that the addition of the capacitor in the base circuit makes the transistor highly susceptible to noise spikes that will be coupled directly to the base of the transistor. Also, capacitors require relatively large areas in integrated circuits; thus RCTL is not the most convenient or popular form of logic for IC fabrication.

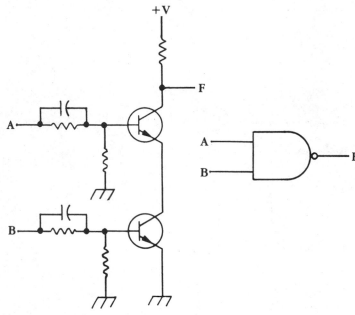

5-10   RCTL NAND gate.

TRANSISTOR-TRANSISTOR LOGIC

With the advent of IC technology, it became feasible to fabricate multiple-emitter transistors. These transistors take the place of diodes in DTL to form a logic family referred to as *transistor-transistor logic* (TTL or T²L). The TTL family combines high-speed operation with single-power-supply capability to make it one of the most popular types of logic. Two configurations of the TTL NAND gate are shown in FIGS. 5-11 and 5-12.

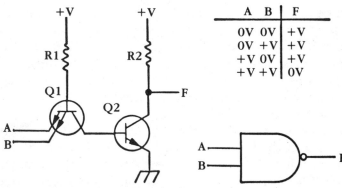

A	B	F
0V	0V	+V
0V	+V	+V
+V	0V	+V
+V	+V	0V

5-11   Basic TTL NAND gate configuration.

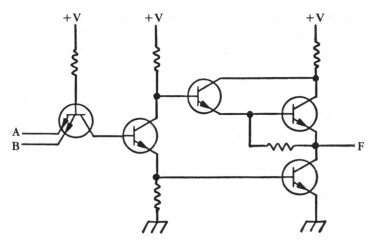

5-12 TTL NAND gate with totem pole output.

In the basic TTL NAND gate configuration, a low potential at either emitter will cause the emitter-base junction of transistor Q1 to be forward-biased, presenting a ground at the base of transistor Q2. Because Q2 is a standard inverter, its output will be nearly + V when ground is present at the base. When both emitters are near + V (input = high), transistor Q1 is cut off and a positive voltage is present at the base of Q2, causing Q2 to conduct. Since Q2 is conducting heavily, + V will be dropped across R2 and the output at F will be 0V.

Although the basic TTL circuit is quite useful, the noise immunity of the configuration is not as good as DTL circuits because of its inherently higher speed switching characteristics. Also, when the output transistor is on, it represents a low-impedance current source to other circuits. However, when the output is switched off, the output reverts to a high-impedance voltage source. This high-impedance state is particularly vulnerable to the pickup of noise transients.

To overcome these problems, a totem-pole output configuration is normally used. Basically, the totem-pole output provides a low-impedance output whether the output state is a logic 1 or a logic 0. This configuration results in very high speed operation with good noise immunity. A typical TTL operating speed is 1 megahertz with propagation delays of approximately 20 nanoseconds.

One disadvantage to the use of the totem-pole output stage is the loss of the ability to connect outputs together. This is often done in DTL and RTL circuits as a means of performing additional gating and is referred to as *collector ORing,* or *wired OR.* If two totem-pole gates were to be wired together, as soon as one of the gates was in the logic 1 state and the other in the logic 0 state, there would be a low-impedance path from the supply voltage to ground, thus inhibiting normal circuit operation and permitting one or both of the output transistors to draw excessive current. Therefore, when one or more outputs must be connected together, two alternatives are available.

The simplest method is through the use of open collector gates. These gates are simply the basic TTL NAND configuration with resistor R2 not provided. One common collector resistor is connected externally, and as many lines as desired connected together. The disadvantage to this scheme is, of course, loss of the totem-pole configuration that provided the speed and drive capability.

The second approach is through the use of three-state TTL gates. The term *three-state* refers to the fact that there are three possible conditions the output can assume: a low-impedance logic 0 state, a low-impedance logic 1 state, or a high-impedance state where the output is considered essentially open.

As can be seen from FIG. 5-13, when the inhibit input is a low, the extra input emitter is in the high state and the circuit acts identically to the stan-

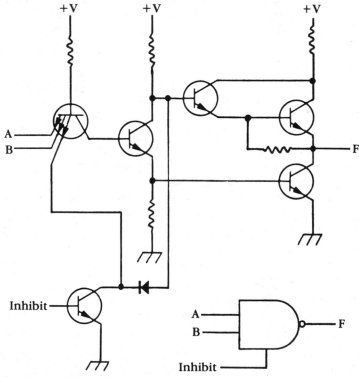

5-13   Three-state TTL NAND gate.

dard totem-pole configuration. However, if the inhibit input is set high, both totem-pole output transistors are driven to their cutoff states and the output of the circuit acts as an open circuit to other similar gates that might be connected on the same line. The only precaution that must be taken with this circuit is that only one three-state gate can be enabled at a time; otherwise, the same effects would occur as with a standard totem-pole output.

EMITTER-COUPLED LOGIC

Thus far, all of the transistor logic circuits described have been classified as saturated-mode switching circuits. That is, the transistors are turned on by driving them into saturation. However, whenever transistors must be switched quickly out of saturation, storage delay time becomes a limiting factor in determining operating speeds.

Storage delay time can be eliminated by operating transistors in a current mode, as shown in the simplified circuit of FIG. 5-14. Logic that operates in this mode is sometimes called *current-mode logic* (CML).

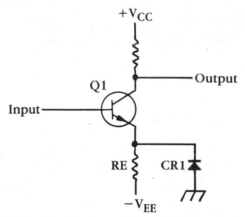

5-14 Simplified current-mode ECL circuit.

When transistor Q1 in the figure is cut off, diode CR1 is forward biased and the emitter of Q1 is held at approximately − 0.7V. A negative voltage is required at the base of Q1 if cutoff is to be maintained. If a slightly positive voltage is applied to Q1, thus turning it on, diode CR1 becomes reverse biased and current flows through the transistor and through R_E. Provided that V_{EE} is much larger than the positive voltage applied at the base of Q1, the current through R_E is essentially the same as when diode CR1 was conducting.

It can be seen that a constant current is maintained in the emitter resistor and that the transistor never goes into saturation. Hence, the current mode of operation allows very fast switching times by eliminating saturated-mode operation.

A handier form of the current-mode circuit is shown in FIG. 5-15. A transistor has been added in place of the diode, and two input transistors are connected in parallel to permit logic gating. The two-transistor combination of Q2 and Q3 is a differential amplifier pair that provides complementary outputs at the two collectors. Because of the common emitter resistor used in the differential amplifier, the circuit is referred to as emitter-coupled; logic that utilizes this configuration is often referred to as *emitter-coupled logic* (ECL). Thus, CML and ECL are two different names for the same type of circuit.

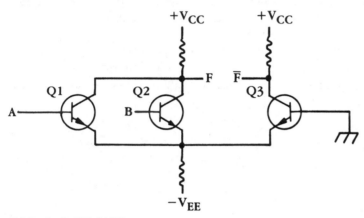

5-15   Basic ECL NOR gate.

In the schematic shown, if both inputs A and B are slightly negative, transistors Q1 and Q2 will be cut off and the emitter-base junction of Q3 will become forward biased, causing Q3 to conduct. If either of the inputs goes slightly positive, the associated transistor (Q1 or Q2) will conduct, cutting off transistor Q3.

One problem with ECL is that it is a low-level logic. This factor makes ECL quite noise susceptible, and great care must be taken to reduce noise spikes and ground noise. However, the extra care required may well be worthwhile where the high operating speeds of ECL are necessary. Disadvantages to ECL are that multiple power supplies are required, and that ECL does not readily interface with other logic families without special buffering.

METAL-OXIDE SEMICONDUCTOR LOGIC

MOS logic utilizes field-effect transistors (FETs) in place of conventional bipolar transistors to form gates and delay elements. Recall that the FET operates by using an electric field to control current flow in a conducting channel. The current in the channel is derived from the flow of majority carriers—holes for p-channel MOS or electrons for n-channel MOS.

Early versions of the insulated-gate field-effect transistor used a metal-oxide-semiconductor layered construction similar to the one shown in FIG. 5-16 (top). The metal portion was aluminum, used to form the gate, which was separated from a silicon semiconductor channel by an insulating layer of silicon dioxide. Today, however, the term is not limited to this specific construction and refers to any insulated-gate field-effect transistor.

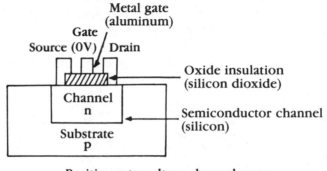

Positive gate voltage-channel open

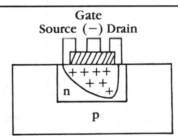

Negative gate voltage-channel depeleted (closed)

5-16   Depletion-mode MOS device.

There are two basic MOS structures that should be understood: depletion-mode MOS devices, and enhancement-mode MOS devices. Both are generally termed *MOSFETs*.

The depletion-mode MOS structure is shown in FIG. 5-16 (bottom). A physical channel of n or p material is diffused into a lightly doped substrate material. At one end is a drain terminal. The flow of current between source and drain is modulated by a voltage applied to a gate terminal, which is insulated from the source and drain. When the gate voltage is at ground potential, the insulation acts as a capacitor dielectric to cause the n material to be basically uncharged. As a result, electrons flow freely through the channel. However, when the gate voltage is made negative, the capacitive effect of the gate plus insulation is to induce a positive potential into the channel, causing depletion of electrons in the n material. As a result, the channel is "pinched off" so that current flow is reduced or cut off. The positive charge in the n channel is effectively a depletion of electrons, the majority carriers, hence the term *depletion-mode MOS.*

The enhancement-mode MOS device is shown in FIG. 5-17. This device does not actually have a physical channel, but instead has two n-type wells at either end of the gate structure. In this case, the gate can be modulated to induce a channel between the two n-type wells, this can be considered an enhancement of electrons in the channel, the device is referred to as a enhancement-mode MOS. Assuming the quiescent condition where the gate is 0V, the p-type substrate material does not permit current flow between the two n-type wells, and the channel is effectively closed. But when a positive gate voltage is applied, the gate capacitive effect induces a negative potential in the p-type substrate, thereby inducing a channel and

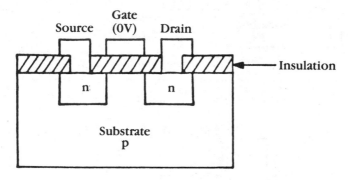

Negative gate voltage-channel closed

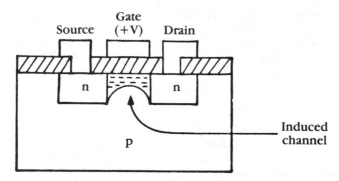

Positive gate voltage-channel enhanced (open)

5-17 Enhancement-mode MOS device.

supplying electrons for current flow between the two wells. Figure 5-18 shows MOS schematic symbols.

Symbol for n-channel MOS   **Symbol for p-channel MOS**

B Active bulk (substrate)
D Drain
G Gate
S Source

5-18   Schematic symbols for MOS devices.

The most obvious difference between the enhancement mode and depletion mode devices is that in the quiescent condition (no gate voltage), current flows in the depletion-mode device but not in the enhancement-mode device. For this reason, enhancement-mode MOS transistors are used in most applications. Both examples shown were for n-channel MOS devices. However, p-channel MOS operates in a similar manner except that the p-channel MOS requires opposite polarity gate voltages for the same conditions.

The operation of MOS logic gates can be explained easily from the schematic diagrams for either p or n MOS circuits. Because the main differences are in the polarity of the supply voltages, the following description will be provided only for p-channel MOS circuits.

A typical MOS inverter circuit is shown as part of FIG. 5-19. One of the most apparent differences between this circuit and the circuit of a bipolar transistor inverter is the substitution of a transistor in place of a resistor as a load.

Basically, MOS devices are integrated circuits that are fabricated on silicon chips. A resistor on one of these chips requires a fairly large area, while a transistor can be fabricated in a small area. For this reason, a transistor normally serves as the load. The load transistor is held conducting at all times by maintaining its gate voltage $(-V_G)$ slightly more negative than the drain voltage $(-V_D)$.

When the input signal is at 0V, the MOS transistor is off and the output is approximately $-V_G$. Provided that the gate voltage is close to the drain voltage, the voltage drop across the load transistor is very small. When the input signal goes negative, the input transistor is turned on and the output falls to ground potential. In this case, approximately V_D is dropped across the load transistor.

Operation of NAND and NOR gates is quite similar to the basic inverter

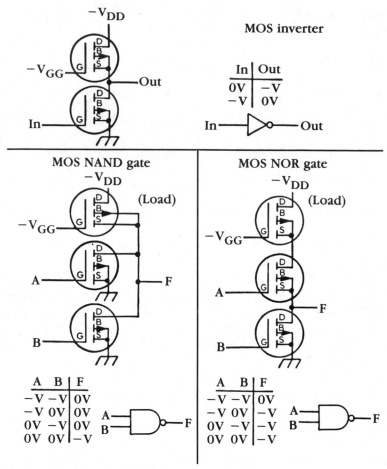

5-19   Examples of p-MOS logic gates.

operation. The NAND gate has two transistors in parallel, while the NOR gate has two transistors in series. Remember that since p-channel MOS uses negative supply voltages, a logic 1 is often assigned as the negative voltage and a logic 0 as ground.

When using negative-true logic in this manner, the gates just described serve opposite functions from those noted. That is, the NAND gate in negative-true logic becomes a NOR gate, and vice versa.

The main advantages to MOS logic are its high noise immunity due to large logic voltage swings and, as a result, the ability to use unregulated power supplies, which are considerably cheaper than regulated supplies. Another significant advantage to MOS logic is that very high packaging densities compared to bipolar speeds can be obtained than for other types of logic, and that it is difficult to interface MOS with other logic families.

COMPLEMENTARY MOS

Complementary MOS (CMOS) circuits utilize both n-MOS and p-MOS transistors on a single chip to form circuits with very low power dissipation. Further, CMOS circuits are typically operated from a single power supply, instead of two supplies as are used with regular MOS.

A typical CMOS inverter circuit is shown in FIG. 5-20. The inverter

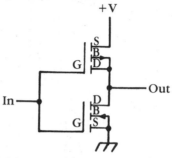

5-20 CMOS inverter.

consists of an n-channel MOS transistor in series with a p-channel MOS transistors. When the input to the inverter is 0V, the n-MOS transistor is off and the p-MOS transistor is on. Thus, the output of the inverter is + V for a 0V input. When the input switches to + V, the n-MOS transistor turns on and the p-MOS transistor switches off. In this case, the output goes to 0V. It can be seen that the output in each case is inverted from the input. No matter which state the input assumes, one MOS transistor is on and the other is off. As a result, steady-state power consumption for CMOS circuits is very low, and the only time that large amounts of current are drawn is during actual switching operations—the transition from off to on.

A CMOS NOR gate is formed by connecting two n-MOS transistors in parallel and two p-MOS transistors in series, as shown in FIG. 5-21. When both of the inputs, A and B, are at 0V the two p-MOS transistors are on and the two n-MOS transistors are off. Thus, + V is available at the output, F. When either input changes to + V, the associated n-MOS transistor turns on, completing a path to ground. Then, the p-MOS transistor opens, disconnecting + V from the output. The resultant output is therefore 0V for any + V input.

Similarly, a CMOS NAND gate is formed by connecting two n-MOS transistors in series and two p-MOS transistors in parallel. If either input goes to 0V, the associated p-MOS transistor is turned on and the series path to ground is opened by the n-MOS transistor turning off. Thus, the output goes to + V. When both inputs are at + V, the p-MOS transistors are turned on, connecting a ground to the output.

CMOS NOR gate

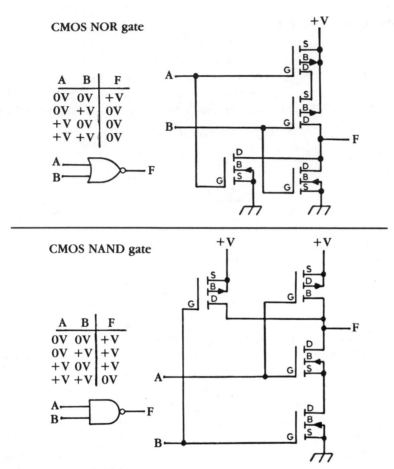

A	B	F
0V	0V	+V
0V	+V	0V
+V	0V	0V
+V	+V	0V

CMOS NAND gate

A	B	F
0V	0V	+V
0V	+V	+V
+V	0V	+V
+V	+V	0V

5-21   Typical CMOS gate structures.

Summary

Digital logic may be implemented in a number of different ways, using different combinations of relays, resistors, diodes, transistors, capacitors, and MOSFETs. Each logic family has different characteristics, which make each of them especially suitable for a particular application. Important considerations are physical size, voltage requirements, current-carrying capabilities, switching speed, noise immunity, and power dissipation.

Questions

1. For each of the following logic families, state the characteristics of construction and the advantages and disadvantages: **a.** relay **b.** diode **c.** diode-transistor **d.** direct-coupled transistor **e.** resistor-transistor **f.** resistor-capacitor-transistor **g.** transistor-transistor **h.** emitter-coupled **i.** metal-oxide semiconductor **j.** complementary MOS

2. What is a high-impedance state? Why is it important?

3. What are the two different types of MOSFETs? How are they similar? How are they different?

 ## Problems

1. Using relay logic, construct a two-input NAND gate.

2. Using relay logic, construct a two-input NOR gate.

3. Using diode logic, construct a circuit for the Boolean expression A OR B OR $CD = F$.

4. Using diode logic, construct a circuit for the Boolean expression AB OR AC OR $BC = F$.

5. Using diode-transistor logic, construct an OR gate.

6. Using diode-transistor logic, construct an AND gate.

7. For the four possible input combinations in the top part of FIG. 5-9, show how the output is determined.

8. Repeat Problem 7 for the bottom part of FIG. 5-9.

9. Draw a diagram for a CMOS OR gate.

10. Draw a diagram for a CMOS AND gate.

BOOLEAN ALGEBRA

AFTER YOU COMPLETE this chapter, you will be able to:

☐ Discuss the laws of Boolean algebra
☐ Understand DeMorgan's theorems
☐ Describe sum-of-product and product-of-sum forms
☐ Explain Karnaugh mapping, which is used to simplify logical expressions

Chapter 4 introduced the concept of Boolean algebra to initiate the presentation of the various logic functions. This chapter contains a brief review of the operating concepts, followed by variations and uses of this mathematical manipulation technique.

Boolean algebra is a set of rules, laws, and theorems by which logical operations can be expressed symbolically in equation form and be manipulated mathematically. Its applications are a convenient and systematic way of expressing and analyzing the operations of digital circuits and systems. In 1854, George Boole published a classic book entitled *An Investigation of the Laws of Thought on Which Are Founded the Mathematical Theories of Logic and Probabilities,* which contained the basics of today's Boolean algebra.

Essentially, any circuit consisting of switch and relay combinations could be represented by mathematical expressions. Today, of course, semiconductor circuits have, for the most part, replaced mechanical switches and relays. However, the same logical analysis is still valid, and a basic knowledge in this area is essential to the study of digital logic.

REVIEW

In most applications of Boolean algebra and in this book, the following concepts are used as a basis for all operations.

- Letters represent variables and functions of variables: e.g., *A, B, C, F,* etc.
- Any single variable or function of a variable has a value of either 1 or 0.

- As explained in chapter 1, positive logic is used; that is, a binary 1 represents a high and a 0 represents a low.

- The complement of a variable is designated by an overscore or *bar* over the letter(s). For example, the variable A is complemented (inverted) with $\overline{A}$. If A = 0, then $\overline{A}$ = 1. Such a complement ($\overline{A}$) is read "A not."

- The logical OR function is represented with a plus sign between the values to be ORed. (See FIG. 6-1.) The OR function of binary numbers is summarized as follows:

$$0 + 0 = 0$$
$$0 + 1 = 1$$
$$1 + 0 = 1$$
$$1 + 1 = 1$$

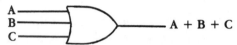

6-1   Variables to be ORed are joined with a plus sign.

- The logical AND function is represented with a dot between the variables to be ANDed (or the dot can be omitted, with the variables side by side), as in regular mathematics (i.e., $A \cdot B$ or AB). The latter is frequently used due to its simplicity (See FIG. 6–2.) The AND function of binary numbers is summarized as follows:

$$0 \cdot 0 = 0$$
$$0 \cdot 1 = 0$$
$$1 \cdot 0 = 0$$
$$1 \cdot 1 = 1$$

6-2   Variables to be ANDed are usually simply placed next to one another.

LAWS OF BOOLEAN ALGEBRA

As in other areas of mathematics, certain well-developed laws and rules must be followed to properly apply Boolean algebra. The three most important laws are: the commutative law, the associative law, and the distributive law.

Commutative Law

The commutative law of addition for two variables is written algebraically as:

$$A + B = B + A$$

This law states that the order in which the variables are ORed makes no difference. Remember that in Boolean algebra terminology as applied to digital circuits, addition and the OR function are the same. Figure 6-3 illustrates the commutative law as applied to the OR gate.

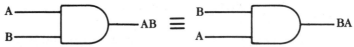

6-3.   The commutative law as applied to an OR gate.

The commutative law of multiplication of two variables is:

$$AB = BA$$

This law states that the order in which the variables are ANDed makes no difference. FIGURE 6-4 illustrates this law as applied to the AND gate.

6-4.   The commutative law as applied to an AND gate.

Associative Laws

The associative law of addition is stated as follows for three variables:

$$A + (B + C) = (A + B) + C$$

This law states that in the ORing of several variables, the result is the same regardless of the grouping of the variables. Figure 6-5 illustrates this law as applied to OR gates.

The associative law of multiplication is stated as follows for three variables:

$$A(BC) = (AB)C$$

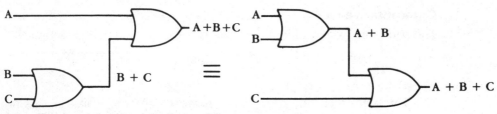

6-5.　The associative law as applied to OR gates.

This law states that it makes no difference in what order the variables are grouped when ANDing several variables. Figure 6-6 illustrates this law as applied to AND gates.

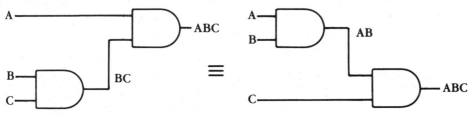

6-6.　The associative law as applied to AND gates.

Distributive Law

The distributive law is written for three variables as follows:

$$A(B + C) = AB + AC$$

This law states that ORing several variables and ANDing the result with a single variable is equivalent to ANDing the single variable with each of the several variables and ORing the products.

This law and the ones previously discussed should be familiar because they are the same as in ordinary algebra. Keep in mind that each of these laws can be extended to include any number of variables. Figure 6-7 illustrates the distributive law in terms of gate implementation.

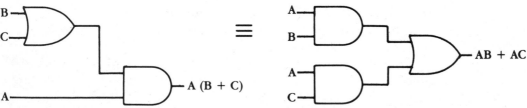

6-7.　The distributive law in terms of gate implementation.

RULES FOR BOOLEAN ALGEBRA

Table 6-1 lists several basic rules that are useful in manipulating and simplifying Boolean expressions. Consider rules 1 through 9 of the table in terms of their application to logic gates. Rules 10 through 12 are derived from the simpler rules and laws covered in this chapter.

Table 6-1. Rules for Boolean Algebra

1. $A + 0 = A$
2. $A + 1 = 1$
3. $A \cdot 0 = 0$
4. $A \cdot 1 = A$
5. $A + A = A$
6. $A + \overline{A} = 1$
7. $A \cdot A = A$
8. $\underline{A} \cdot \overline{A} = 0$
9. $\overline{\overline{A}} = A$
10. $A + AB = A$
11. $A + \overline{A}B = A + B$
12. $(A + B)(A + C) = A + BC$

Note: *A* can represent a single variable or a combination of variables.

Rule 1 can be observed by what happens when one input to an OR gate is always 0 and the other input, *A,* can take a 1 or 0 value. If *A* is a 1, the output is 1, which is equal to *A.* If *A* is a 0, the output is 0, which is also equal to *A.* Therefore, a variable ORed with a 0 is equal to the value of the variable $(A + 0 = A)$. This rule is further demonstrated in FIG. 6-8, where input is fixed at 0.

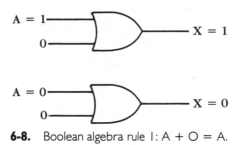

6-8. Boolean algebra rule 1: A + O = A.

Rule 2 is demonstrated when one input to an OR gate is always 1 and the other input, *A,* takes on either a 1 or 0 value. A 1 on an input to an OR gate produces a 1 on the output, regardless of the value of the variable on the other input. Therefore, a variable ORed with a 1 is always equal to 1 ($A + 1 = 1$). This rule is illustrated in FIG. 6-9, where input B is fixed at 1.

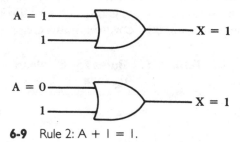

6-9   Rule 2: A + I = I.

Rule 3 is demonstrated when a 0 is ANDed with a variable. Recall that anytime one input to an AND gate is 0, the output is 0, regardless of the value of the variable on the other input. A variable ANDed with a 0 always produces a 0 ($A \cdot 0 = 0$). This rule is illustrated in FIG. 6-10, where input B is fixed at 0.

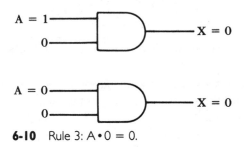

6-10   Rule 3: A • 0 = 0.

Rule 4 can be verified by ANDing a variable with a 1. If the variable *A* is a 0, the output of the AND gate is a 0. If the variable is a 1, the output of the AND gate is a 1 because both inputs are now 1's. Therefore, the results of ANDing a variable and 1 is the value of the variable ($A \cdot 1 = A$). This rule is shown in FIG. 6-11, where input B is fixed at 1.

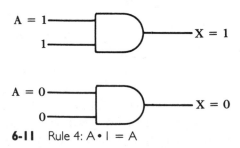

6-11   Rule 4: A • I = A

Rule 5 states that if a variable is ORed with itself, the output is equal to the variable. For instance, if A is a 0, then $0 + 0 = 0$, and if A is a 1, then $1 + 1 = 1$. Figure 6-12 illustrates this rule.

6-12 Rule 5: $A + A = A$.

Rule 6 can be explained as follows: if a variable and its complement are ORed, the result is always a 1. If A is a 0, then $0 + 0 = 0 + 1 = 1$. If A is a 1, then $1 + 1 = 1 + 0 = 1$. See FIG. 6-13 for an illustration of this rule.

6-13 Rule 6: $A + \overline{A} = 1$.

Rule 7 states that if a variable is ANDed with itself, the result is equal to the variable. For example, if $A = 0$, then $0 \cdot 0 = 0$, and if $A = 1$, then $1 \cdot 1 = 1$. For either case, the output of an AND gate is equal to the value of the input variable A. Figure 6-14 illustrates this rule.

6-14 Rule 7: $A \cdot A = A$.

Rule 8 states that if a variable is ANDed with its complement, the result is 0. This is readily apparent because either A or $\overline{A}$ will always be 0, and when a 0 is one of the inputs to an AND gate, the output is always 0. Figure 6-15 illustrates this rule.

6-15   Rule 8: $A \cdot \overline{A} = 0$.

Rule 9 simply states that if a variable is complemented twice, the result is the variable itself. Starting with A and inverting (complementing) it once gives $\overline{A}$. Inverting it once more gives A—the original value. This rule is illustrated in FIG. 6-16.

6-16   Rule 9: $\overline{\overline{A}} = A$.

Rule 10 can be proven by using the distributive law, rule 2, and rule 4 as follows:

Rule 10: $A + AB = A$

$$
\begin{aligned}
A + AB &= A(1 + B) & \text{distributive law} \\
&= A \cdot 1 & \text{rule 2} \\
&= A & \text{rule 4}
\end{aligned}
$$

Rule 11 can be proven as follows:

$$A + \overline{A}B = A + B$$

$$
\begin{aligned}
A + AB &= (A + AB + AB & \text{rule 10} \\
&= (AA + AB) + \bar{A}B & \text{rule 7} \\
&= AA + AB + A\bar{A} + \bar{A}B & \text{rule 8 (adding AA = 0)} \\
&= (A + \bar{A})(A + B) & \text{by factoring}
\end{aligned}
$$

$$= 1(A + B) \qquad \text{rule 6}$$
$$= A + B \qquad \text{rule 4}$$

Rule 12 can be proven as follows:

$$(A + B)(A + C) = A + BC$$

$$
\begin{aligned}
(A + B)(A + C) &= AA + AC + AB + BC &&\text{distributive law} \\
&= A + AC + AB + BC &&\text{rule 7} \\
&= A(1 + C) + AB + BC &&\text{distributive law} \\
& &&\text{and rule 4} \\
&= A{\cdot}1 + AB + BC &&\text{rule 2} \\
&= A(1 + B) + BC &&\text{distributive law} \\
&= A{\cdot}1 + BC &&\text{rule 2} \\
&= A + BC &&\text{rule 4}
\end{aligned}
$$

DEMORGAN'S THEOREMS

DeMorgan was a logician and mathematician who was acquainted with Boole. He proposed two theorems that are an important part of Boolean algebra. In equation form, they are:

$$\overline{AB} = \overline{A} + \overline{B}$$
$$\overline{A + B} = \overline{A}\,\overline{B}$$

The first theorem can be stated as follows: *the complement of a product is equal to the sum of the complements.* What it is actually saying is that the complement of two or more variables ANDed is the same as the OR of the complements of each individual variable.

The second theorem can be stated as follows: *the complement of a sum is equal to the product of the complements.* This is stating that the complement of two or more variables ORed is the same as the AND of the complements of each individual variable. These theorems are illustrated in FIG. 6-17.

BOOLEAN EXPRESSIONS FOR GATE NETWORKS

As you can see from the previous examples, the form of a Boolean expression indicates the type of gate network it describes. For example, the logic diagram in FIG. 6-18 illustrates the gate network that the expression A(B + CB) represents.

First, there are four variables in the figure: *A, B, C,* and *D.* Variable *C* is ANDed with *D,* resulting in *CD.* Then, *CD* is ORed with *B,* giving (*B + CD*). This sum is then ANDed with *A* to produce the final function.

A	B	$\overline{AB}$	$\overline{A}+\overline{B}$
0	0	1	1
0	1	1	1
1	0	1	1
1	1	0	0

A	B	$\overline{A+B}$	$\overline{A}\overline{B}$
0	0	1	1
0	1	0	0
1	0	0	0
1	1	0	0

For $\overline{AB} = \overline{A} + \overline{B}$

For $\overline{A + B} = \overline{A}\overline{B}$:

6-17 DeMorgan's theorems.

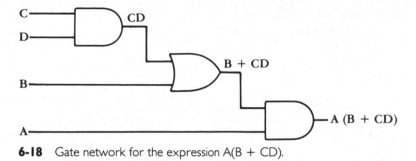

6-18 Gate network for the expression A(B + CD).

Therefore, the form of the expression determines how many and which types of logic gates are needed, as well as how they are connected together. The more complicated the expression, the more complex the gate network will be. It is therefore best to simplify an expression as much as possible to get the simplest gate network. (Simplification methods are in the next section.) Two common forms for Boolean expressions that can be helpful in simplifying gate networks are the *sum-of-products* form and the *product-of-sums* form.

Sum-of-Products Form

The sum-of-products form of a Boolean expression is the form in which two or more products (variables that are ANDed) are added (ORed). For example, $AB + CD$ is a sum-of-products expression. Other examples of these expressions are:

$$A\overline{B}C + DEF + A\overline{EF}$$
$$\overline{A} + BC\overline{D} + EFG$$

One reason the sum-of-products is a useful form of a Boolean expression is the straightforward manner in which it can be implemented in logic gates. As FIG. 6-19 shows, it is simply two steps: ANDing and then ORing. Therefore, this form is always only a *two-level* gate network; that is, the maximum number of gates through which a signal must pass in going from input to output is two (excluding inversions, but these can also be worked in).

Product-of-Sums Form

The product-of-sums form is the ANDing of two or more sums (variables that are ORed). For instance, $(A + B)(C + D)$ is a product of sums. This form also lends itself to simpler gate networking, as was the sum of products, in that a two-level network is also formed. (See FIG. 6-20.)

Expression: AB + BCD + EF

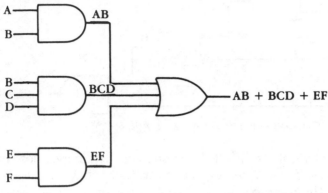

6-19   Gate network that results in a sum-of-products form.

Expression: (A + B) (C + D + E) (F + G + H)

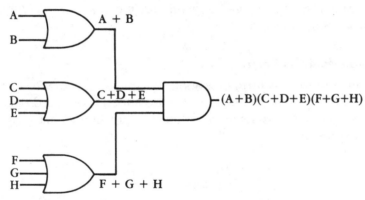

6-20   Gate network that results in a product-of-sums form.

SIMPLIFICATION OF BOOLEAN EXPRESSIONS

In the design of digital circuits, it is necessary to reduce the expression to its simplest form to facilitate the building of the gate network with the least number of components. Hence, this section focuses on the basic laws, rules, and theorems of Boolean algebra and other methods needed to manipulate and simplify expressions.

As an example, consider the expression $AB + A(B + C) + B(B + C)$ and simplify it using the Boolean algebraic techniques discussed thus far. This is one possible solution:

Step 1: Apply the distributive law to the second and third terms in the expression as follows:

$$AB + AB + AC + BB + BC$$

Step 2: Apply rule 7 ($BB = B$):

$$AB + AB + AC + B + BC$$

Step 3: Apply rule 5 ($AB + AB = AB$):

$$AB + AC + B + BC$$

Step 4: Factor B out of the last two terms:

$$AB + AC + B(1 + C)$$

Step 5: Apply the commutative law and rule 2 ($1 + C = 1$):

$$AB + AC + B \cdot 1$$

Step 6: Apply rule 4 ($B \cdot 1 = B$):

$$AB + AC + B$$

Step 7: Factor B out of the first two terms:

$$B(A + 1) + AC$$

Step 8: Apply rule 2 ($A + 1 = 1$):

$$B \cdot 1 + AC$$

Step 9: Apply rule 4 ($B \cdot 1 = B$)

$$B + AC$$

The expression is now simplified as far as it can go. Note that once you get acquainted with Boolean simplification techniques, you can combine many individual steps.

Karnaugh Map Simplification

Another approach for reducing a Boolean expression to its simplest or minimum form is known as the *Karnaugh map method,* named for its originator. The method is systematic and easily applied. When properly used, it always results in the minimum expression possible.

The effectiveness of algebraic simplification as expressed in the previous sections is dependent on your ability and ingenuity in applying the rules, laws, and theorems of Boolean algebra. The Karnaugh map approach, therefore, has a distinct advantage, especially for more complex expressions where algebraic simplification is not immediately obvious or is extremely involved.

A Karnaugh map is composed of a number of adjacent "cells." Each cell represents one particular combination of variable values. Since the total number of possible combinations of n variables is 2^n, there must be 2^n cells in the Karnaugh map.

Karnaugh maps are representations of truth tables that have been organized into cells. Each cell represents a 1-bit change in the input variables when compared to the next adjacent cell. The truth table for a NOR function is shown in FIG. 6-21 as an example of a two-variable function that can be transferred to a Karnaugh map. In this map, a 1 is placed in the cell corresponding to the $\overline{A}\overline{B}(00)$ state, while a 0 is placed in each of the other three cells. The equation representing the function shown is $\overline{AB} = F.$

A three-variable Karnaugh map can be formed from a three-variable truth table in a similar manner. Upon inspection of the truth table in the figure, it might be thought that a reasonable equation to represent the given function is $\overline{A}BC + \overline{A}\overline{B}C = F.$ However, the Karnaugh map shows that an output of 1 results no matter which state variable B assumes and that the only requirement to fulfill the function is that $\overline{A}C = F.$

Of course, all this can be verified or derived by manipulation of the logic equations. As a general rule, however, algebraic manipulation is more helpful in expressing a function in some other form, while the Karnaugh map is useful for obtaining the simplest possible expression of the truth table.

A final example in FIG. 6-21 illustrates the application of a Karnaugh map to a four-variable function. Here, the truth table shows that a logic 1 output is obtained for four different conditions. It might not be clear from the truth table that each of these conditions does not require a four-input gate. Reference to the Karnaugh map, however, makes it clear that all the logic 1 outputs are obtained when C and D are both a logic 1 and that the states of A and B do not matter. Thus, an equation that adequately represents this function is $CD = F,$ which can be implemented with a single two-input NAND gate.

Using the Karnaugh Map

Various groups of logic 1's in a Karnaugh map can be considered together, and a resultant simplification of the required logic implementation can be

A	B	F
0	0	1
0	1	0
1	0	0
1	1	0

Two-variable map

$\overline{AB}$ = F

A＼B	0	1
0	1	0
1	0	0

A	B	C	F
0	0	0	0
0	0	1	1
0	1	0	0
0	1	1	1
1	0	0	0
1	0	1	0
1	1	0	0
1	1	1	0

Three-variable map

$\overline{A}$ C = F

AB＼C	0	1
00	0	1
01	0	1
11	0	0
10	0	0

A	B	C	D	F
0	0	0	0	0
0	0	0	1	0
0	0	1	0	0
0	0	1	1	1
0	1	0	0	0
0	1	0	1	0
0	1	1	0	0
0	1	1	1	1
1	0	0	0	0
1	0	0	1	0
1	0	1	0	0
1	0	1	1	1
1	1	0	0	0
1	1	0	1	0
1	1	1	0	0
1	1	1	1	1

Four- variable map

CD = F

AB＼CD	00	01	11	10
00	0	0	1	0
01	0	0	1	0
11	0	0	1	0
10	0	0	1	0

6-21   Basic Karnaugh map configurations.

obtained. Further, it should be clear that even if a truth table does not exist, one can be generated from logic equations or from a logic diagram.

From the truth table, transfer the logic 1 outputs to a Karnaugh map. All that remains is to determine which groupings of logic 1's can be considered together to help in writing a simplified logic equation.

All the combinations of allowed groupings, called *loops,* are shown in FIGS. 6-22 and 6-23 for three- and four-variable maps, respectively. Each loop in a map consists of a single AND function, and all the AND functions that are represented by the various loops are ORed together to obtain the final logic network. Since this method requires one AND function for each loop, a minimum number of loops should be used. Additionally, it is always desirable to draw the largest possible loops, because the larger the loop, the fewer the number of variables contained in each AND function. Translated into logic gates, the smaller the resultant logic equation, the fewer the number of gates required to implement the equation.

As an example of plotting a Boolean function on a Karnaugh map, consider the function $F = \overline{A}BC + AB(\overline{C} + C)$. Place a 1 in each cell corresponding to the terms that make the entire function a 1. To do this, it is very helpful to get the expression in sum-of-products form. For example, the expression $F = \overline{A}BC + AB(\overline{C} + C)$ converted to sum-of-products form is $F = \overline{A}BC + AB\overline{C} + ABC$. Figure 6-24 shows the three-variable map with 1's corresponding to each of the terms in the expression.

Now the 1's must be factored to reduce the expression to its minimum. This is done by combining the adjacent cells containing 1's (draw a loop around them). *Adjacent cells* in Karnaugh mapping are those that differ by only the value of a single variable. For instance, in the map of FIG. 6-24, the upper left and lower left cells are not adjacent because they have two variables that differ between them. The same is true for the upper right and lower right cells.

Therefore, the rules for factoring 1's on a Karnaugh map are as follows. (See FIG. 6-25.)

1) Combine the 1's appearing in adjacent cells into as large a group as possible, as long as the number of 1's in the group is a power of 2 (2, 4, 8, etc.).

2) Form as many maximum-sized groups as possible until all 1's are included in at least one group. One group can overlap into the next, and groups can have only one 1. Remember the end-around loops that you can form as demonstrated in FIG. 6-23.

3) Each group of 1's creates a minimized product term composed of all variables that have the same value (1 or 0) within the group. If a true variable appears in all the cells in the group, the true form of the variable appears in the product form. If a complemented variable appears in all cells of the group, the complement form of the variable appears in the product term. If both true and complement forms appear within the group, they cancel and are omitted from the product term.

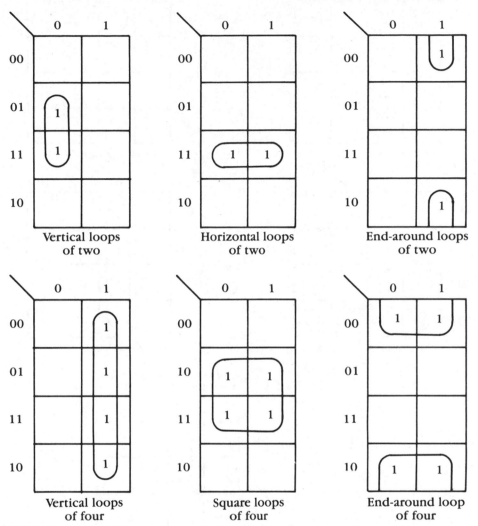

6-22 Three-variable Karnaugh map loops.

4) The product terms represented by each group on the map are then summed (ORed), producing a minimum sum-of-products expression. The expression $F = \overline{A}BC + AB(\overline{C} + C)$ has been simplified to its equivalent $F = AB + BC$.

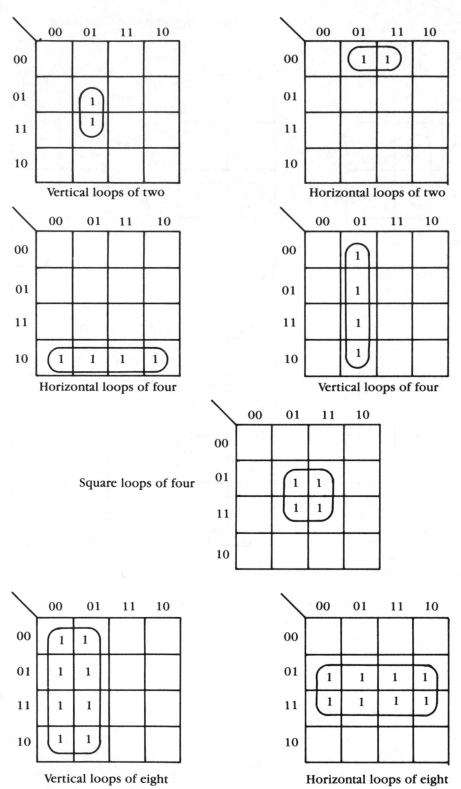

6-23 Four-variable Karnaugh map loops.

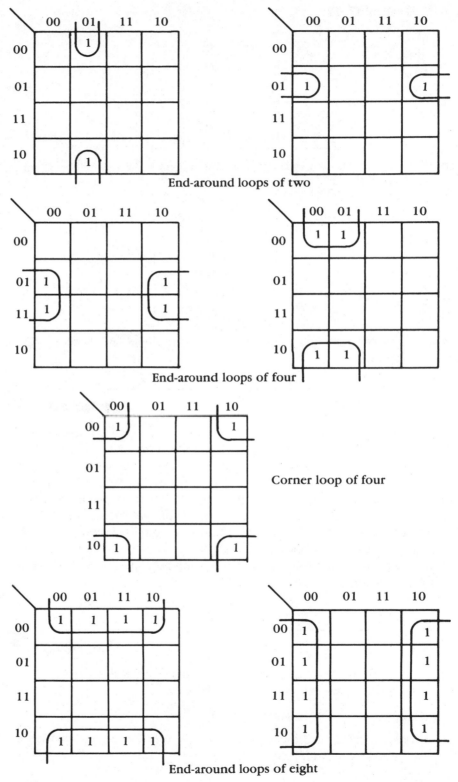

End-around loops of two

End-around loops of four

Corner loop of four

End-around loops of eight

6-23 Continued.

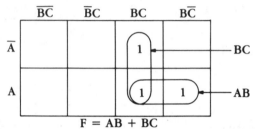

	$\overline{BC}$	$\overline{B}C$	BC	$B\overline{C}$
$\overline{A}$			1	
A			1	1

6-24 Three-variable Karnaugh map for the expression $F = \overline{A}BC + AB(\overline{C} + C)$ (or $F = \overline{A}BC + AB\overline{C} + ABC$).

	$\overline{BC}$	$\overline{B}C$	BC	$B\overline{C}$	
$\overline{A}$			1		— BC
A			1	1	— AB

$$F = AB + BC$$

6-25 Factoring or "grouping" of the 1s for the expression $F = \overline{A}BC + AB\overline{C} + ABC$.

Summary

Three mathematical laws are also used in Boolean algebra. These are the commutative law, the associative law, and the distributive law. There is also a series of rules for simplifying logical expressions, which apply not only to the above laws but to other concepts as well. DeMorgan's theorems are also used for simplification. One theorem is used for converting a NAND statement into an OR statement, and another theorem is used for converting a NOR statement into an AND statement. The two standard forms for a Boolean expression are the sum of products form and the product of sums form. Boolean expressions are also simplified by using Karnaugh maps.

Questions

1. Write logical equations which express each of the following: commutative law, associative law, distributive law.

2. State, in words, DeMorgan's theorems.

3. Write a gate network for the expression $A + B(C + D)$ and convert it to a sum of products form.

4. Write a gate network for the expression $(A + B)(C + D + E)F$.

5. Draw a Karnaugh map for two, three, and four variables. Do not include any values within the map.

Problems

1. Simplify the following expression: $AB + A\overline{C} + C = F$

2. Simplify the following expression: $(A + B)(A + C) = F$

3. Use DeMorgan's theorems to simplify the following expressions:
   a. $\overline{(A+B)} + AB = F$ **b.** $\overline{AB + AC} = F$

4. Use DeMorgan's theorems to simplify the following expressions:
   a. $\overline{A(B+C)} + \overline{(A+C)} = F$ **b.** $\overline{(A+B)(C+D)} = F$

5. For your answers to Problem 3, draw a Karnaugh map.

6. For your answers to Problem 4, draw a Karnaugh map.

7. Draw a Karnaugh map for the expression $A\overline{B} + \overline{A}\,\overline{B}\,\overline{C} + \overline{A}BC = F$ and use it to simplify the expression.

8. Draw a Karnaugh map for the expression $A\overline{B}\,\overline{D} + \overline{A}BC + AB\overline{C}\,\overline{D} = F$ and use it to simplify the expression.

COMBINATIONAL AND SEQUENTIAL LOGIC

AFTER YOU COMPLETE this chapter, you will be able to:

☐ Understand the difference between combinational logic and sequential logic
☐ Calculate the output from a series of gates
☐ Describe gate troubleshooting techniques
☐ Explain the use of timing diagrams
☐ Explain what is meant by a flip-flop
☐ Understand the operation of different types of multivibrators

*P*revious chapters explained the individual gates and how to manipulate the Boolean expressions they represent. This chapter explains how these gates are combined to perform more complex functions.

Logic functions are performed with essentially two types of logic circuits. These approaches are:

• *Combinational Logic*—A network of several gates that are connected to generate a specific output with no storage involved. This type of network *combines* the input variables in such a way that the output is always dependent on the combination of inputs.

• *Sequential Logic*—Logic operations within this type of network occur in a definite *sequence,* providing the ability to store or delay signals.

COMBINATIONAL LOGIC

There are two ways to analyze a combinational network:

• The first is to take a given set of inputs and work through each gate to finally determine the output, or

• Derive a logic equation by analyzing the gates in the combination, plug in the input values, and deduce the output using Boolean algebra and other techniques described in chapter 6.

To demonstrate the first example, FIG. 7-1 shows a combinational logic circuit consisting of two AND gates and one OR gate. Each of the three gates has two input variables as indicated.

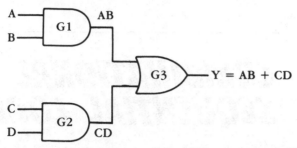

7-1. A combinational logic circuit of two AND gates and one OR gate.

Each of the input variables can be either a high (1) or a low (0). Because there are four input variables, there are 16 possible combinations of the input variables ($2^4 = 16$). To illustrate an analysis procedure, let's assign a possible input combination to see what the corresponding output value is.

First, say each input variable is low. Examine the output of each gate in the network to arrive at the final output Y. If the inputs to gate G1 are both low, the output of G1 is low. Also, the output of G2 is low because both inputs are low, so the two low inputs to G3 also make its output low. Therefore, the output of the logic circuit in FIG. 7-1 is low when all the inputs are low, as shown in TABLE 7-1, which also shows the remaining 15 combinations. Verify each combination in the table from the diagram.

Table 7-1.   Truth Table for Fig. 7-1:
Y = AB + CD

Inputs A B C D	G1 Output (AB)	G2 Output (CD)	G3 Output Y
0 0 0 0	0	0	0
0 0 0 1	0	0	0
0 0 1 0	0	0	0
0 0 1 1	0	1	1
0 1 0 0	0	0	0
0 1 0 1	0	0	0
0 1 1 0	0	0	0
0 1 1 1	0	1	1
1 0 0 0	0	0	0
1 0 1 0	0	0	0
1 0 1 1	0	1	1
1 1 0 0	1	0	1
1 1 0 1	1	0	1
1 1 1 0	1	0	1
1 1 1 1	1	1	1

As a second method of analyzing the operation of the circuit in FIG. 7-1, a logic equation is developed from observing the diagram and, using Boolean algebra, results are obtained for each input combination to generate TABLE 7-1.

Because gate G1 is an AND gate and its two inputs are A and B, its output is expressed as AB. Gate G2 is an AND gate also and its two inputs are C and D, so its output expression is CD. Gate G3 is an OR gate, so its output is the ORing of AB and CD, producing $AB + CD$. Therefore, the output function for the figure is $Y = AB + CD$.

Now simply plug in the values for each combination of inputs and use the Boolean rules and laws to determine the result. For example, when $A = 1, B = 1, C = 1$, and $D = 0$, the result is as follows:

$$Y = AB + CD$$
$$= 1 \cdot 1 + 1 \cdot 0$$
$$= 1 + 0$$
$$= 1$$

AND-OR Invert Logic

Figure 7-2A shows a combinational logic circuit consisting of two AND gates, one OR gate, and an inverter. As you can see, the operation is the same as for the AND-OR circuit in FIG. 7-1 except that the output is inverted. The output expression is $Y = \overline{AB + CD}$. An evaluation of this for the inputs $A = 1, B = 1, C = 1$, and $D = 0$ is as follows:

$$Y = \overline{AB + CD}$$
$$= 1 \cdot 1 + 1 \cdot 0$$
$$= 1 + 0$$
$$= 1$$
$$= 0$$

Notice that DeMorgan's theorem can now be applied to this output expression as follows:

$$Y = \overline{AB + CD}$$
$$= (\overline{A} + \overline{B})(\overline{C} + \overline{D})$$

The resulting equivalent circuit is shown in FIG. 7-2B.

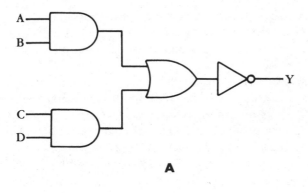

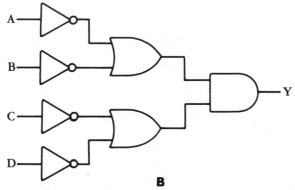

7-2. A combinational logic circuit of two AND gates, one OR gate, and an inverter.

Exclusive-OR Logic

Figure 7-3 shows a combinational logic circuit that does the EX-OR function (recall this type of function back in FIG. 4-7). The output for this circuit is $Y = A\overline{B} + \overline{A}B$. Evaluation of this expression results in the truth table in TABLE 7-2. Notice that the output is high only when the inputs are different.

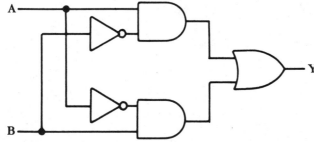

7-3. A combinational logic circuit that performs the EX-OR function.

Table 7-2. EX-OR Truth Table.

A	B	Y
0	0	0
0	0	1
1	0	1
1	1	0

Recalling the EX-OR algebraic symbol, the output expression can also be written as $Y = A \oplus B$, read a "Y equals A exclusive-OR B."

Implementing Combinational Networks

Say you were asked to do the opposite of what was covered in the last section, that is, rather than derive an equation from a series of gates, you want to create the gate network from a given equation.

Consider the following equation:

$$X = AB + CDE$$

The function is composed of two terms, AB and CDE, and it contains a total of five variables. The first term is formed by ANDing A with B, and the second term is formed by ANDing C, D, and E. These two terms are then ORed to form the function X. These operations are indicated in the structure of the equation as follows:

AND (second-level operation)

$$X = AB + CDE$$

OR (first-level operation)

Note that in this particular equation, the AND operations that form the two individual terms AB and CDE must be carried out before the terms can be ORed. The OR operation is the last to be performed to produce the output; therefore, it is called a *first-level operation*, meaning it is performed by the first-level gate, starting at the output and working back to the inputs. The AND gates are completing their function at the second level from the output, so they are second-level operations.

To implement the logic function, a two-input AND gate is needed for the term AB and a three-input gate is required for the term CDE. Then, a two-input OR gate combines the two AND terms. The resulting logic circuit is shown in FIG. 7-4.

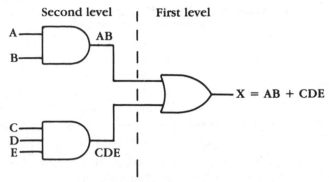

7-4.   The combinational gate network implemented from the expression X = AB + CDE.

As a second example, consider the following equation:

$$X = AB(C\overline{D} + EF)$$

A breakdown of this equation shows that the term AB and the term $C\overline{D} + EF$ are ANDed. The term AB is formed by ANDing the variables A and B. The term $C\overline{D} + EF$ is formed first by ANDing C and $\overline{D}$, ANDing E and F, and then ORing these two terms. This structure is indicated in relation to the equation as follows:

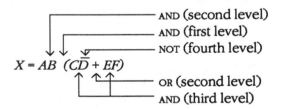

Before the output function X can be performed, the term *AB* must exist, which is formed by ANDing *A* with *B*. The term $C\overline{D} + EF$ must also exist, but before that, the terms $C\overline{D}$ and *EF* must be formed and ORed together, etc. As you can see, there is a chain of logic operations that must be done in the proper order before the output function can be obtained.

The logic gates required to implement $X = AB(C\overline{D} + EF)$ are as follows:

1) One inverter to form $\overline{D}$.

2) Two two-input AND gates to form $C\overline{D}$ and *EF*.

3) One two-input OR gate to form $C\overline{D} + EF$.

4) One two-input AND gate to form *AB*.

5) One two-input AND gate to form *X*.

The logic circuit that performs this function is shown in FIG. 7-5A. Notice that there can be more than one way to implement a given function. For instance, the AND gate in FIG. 7-5A that forms the function *AB* can be eliminated and the inputs *A* and *B* brought into the first-level, three-input AND gate, as shown in FIG. 7-5B. The resulting output is exactly the same, and the circuit used is simpler. Both circuits of FIG. 7-5 have a minimum of four logic levels.

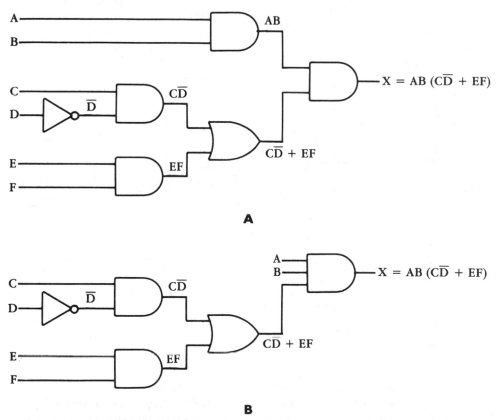

7-5   Logic circuits for the expression X = AB(CD + EF). (A) is the initially derived circuitry, and (B) is the simplified version.

GATE MINIMIZATION

This section shows how to reduce the number of gates required to produce a given function. In many applications, it is desirable to use the minimum number of gates in the simplest configuration possible. The reason might be for economy or cost, limitations in available power, to minimize delay times by reducing the number of logic levels, or to maximize the chip area in the case of IC design.

Here, two basic types of minimizations are examined. First, the rules and laws of Boolean algebra are applied to simplify the equation, and second, the Karnaugh map is used to minimize the function. Recall that both of these methods were described in the last chapter.

First, Boolean algebra is applied to the equation for the circuit in FIG. 7-6. The logic equation for this circuit is

$$X = AC\overline{D} + \overline{A}B(CD + BC)$$

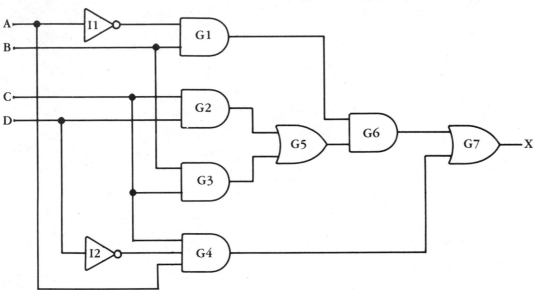

7-6   Circuit for the function $X = AC\overline{D} + \overline{A}B(CD + BC)$.

As the function appears in this equation, five AND gates, two OR gates, and two inverters are required to implement it. Whether this is the simplest form of the equation has yet to be discovered. When using the rules of Boolean algebra to simplify an equation, ingenuity is a must because most equations can be written more than one way and still denote the same function (circuit). The goal is to use all the rules and laws necessary to arrive at what appears to be the simplest expression.

For the equation, a possible first step in simplification is to apply the distributive law to the second term by multiplying the term $CD + BC$ by $\overline{A}B$. The result is

$$X = AC\overline{D} + \overline{A}BCD + \overline{A}BBC$$

The rule that $BB = B$ can be applied to the third term, as

$$X = AC\overline{D} + \overline{A}BCD + \overline{A}BC$$

Notice that C is common to every term, so it can be factored out using the distributive law:

$$X = C(A\overline{D} + \overline{A}BD + \overline{A}B)$$

Now notice that the term $\overline{A}B$ appears in the last two terms within the parentheses and can be factored out of those two terms:

$$X = C[A\overline{D} + \overline{A}B(D + 1)]$$

Since $D + 1 = 1$,

$$X = C(A\overline{D} + \overline{A}B)$$

It appears that this equation cannot be simplified any further, but it can be written in a slightly different way by applying the distributive law (this results in the sum-of-products form):

$$X = AC\overline{D} + \overline{A}BC$$

Implementing this equation into a circuit, it only requires two three-input AND gates, two inverters, and one two-input OR gate, as shown in FIG. 7-7. Compare this circuit to that in FIG. 7-6. The minimized circuit is equivalent, but only requires five gates (instead of nine), and the number of levels has been reduced from five to three, allowing for a shorter propagation delay through the circuit.

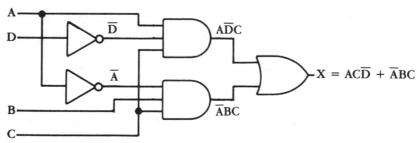

7-7 Simplified diagram from that of Fig. 7-6. The expression is $X = AC\overline{D} + \overline{A}BC$.

Karnaugh Map Approach

The second approach to gate minimization is done using a Karnaugh map, which is more systematic. Further, if factoring is done correctly, the minimum expression will result. For comparison purposes, the same function that was used in the last example is used:

$$X = AC\overline{D} + \overline{A}B(CD + BC)$$

First convert the expression to sum-of-products form:

$$X = AC\overline{D} + \overline{A}BCD + \overline{A}BC$$

Plot the function on the Karnaugh map, as shown in FIG. 7-8. Note that both the first and third terms include two cells on the map: $AC\overline{D}$ includes cells

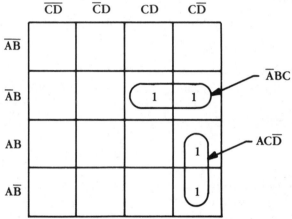

7-8   Karnaugh map for the function of X = AC$\overline{D}$ + $\overline{A}$B(CD + BC).

corresponding to $ABC\overline{D}$ and $A\overline{B}C\overline{D}$. $\overline{A}BC$ includes cells corresponding to $\overline{A}BCD$ and $\overline{A}BC\overline{D}$.

By factoring the map as indicated, the two terms shown are the result, and the function can be expressed in the minimum, sum-of-products form:

$$X = AC\overline{D} + \overline{A}BC$$

This equation is the same as that in the example using Boolean laws and rules. The map method of logic simplification is especially useful when the logic function is long and cumbersome. The application of Boolean rules and laws to extremely complex functions tends to be tedious and more of a trial-and-error process, whereas the map method is very systematic and always yields a minimum result.

TROUBLESHOOTING GATE NETWORKS

In a typical combinational logic network, the output of one gate is connected to two or more gate inputs as shown in FIG. 7-9. In the figure, the

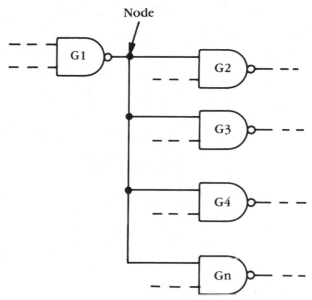

7-9 The output of one gate as connected to the inputs of several others via a node.

topmost node is the driving node (to G1), where the other gates represent loads connected to the node. Several types of failures are possible in this situation. Some of these failures are difficult to isolate down to one bad gate because all of the gates connected to the node are affected. The types of failures discussed in this section are:

Open output in driving gate. This failure causes a loss of signal to all load gates.

Open input in a load gate. This failure does not affect the operation of any other gates connected to the node, but it does result in a loss of signal output from the faulty gate.

Shorted output in driving gate. This failure can cause the node to be stuck in the low state.

Shorted input in a load gate. This failure can also cause the node to be stuck in the low state.

Open Output in a Driving Gate

In this situation, there is no pulse activity on the node. With circuit power on, an open node normally results in a "floating" level and is indicated by a dim lamp on a logic probe. See FIG. 7-10.

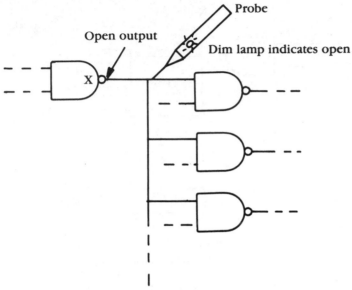

7-10   Circuit condition that results when there is an open output on the driving gate.

Open Input in a Load Gate

If the check for an open driver output is negative, then a check for an open input in a load gate should be done. With circuit power off, apply the logic pulser tip to the node. Then check the output of each gate for pulse activity with the logic probe as illustrated in FIG. 7-11. If one of the inputs that is normally connected to the node is open, no pulses will be detectable on that gate output.

Shorted Output in a Driving Gate

This fault can cause the node to be stuck at a low level. A check with the logic probe can indicate this as shown in FIG. 7-12A. A short to ground in the driving gate's output or in any gate input causes this symptom, and therefore further checks must be made to isolate the short to a particular gate.

If the driving stage's output is internally shorted to ground, then essentially no current activity can be present on any of the connections to the node. Therefore, a current tracer will indicate no activity with circuit

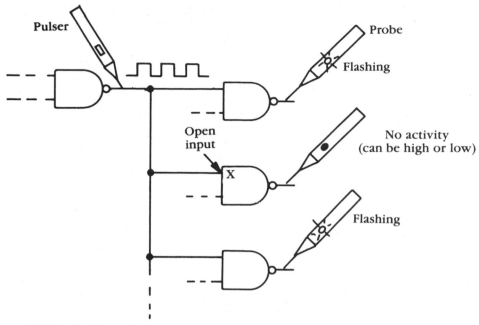

7-11   Circuit condition when there is an open input on a load gate.

power as illustrated in FIG. 7-12B. To further verify a shorted output, a pulser and current tracer can be used with the circuit power off as shown in FIG. 7-12C. When current pulses are applied to the node with the pulser, all of the current will flow into the shorted output and none will flow through the circuit paths into the load gate inputs.

Shorted Input in a Load Gate

If one of the load gate inputs is internally shorted to ground, the node will be stuck in the low state. Again, as in the case of a shorted output, the logic pulser and current tracer can be used to isolate the faulty gate.

When the node is pulsed with the circuit power off, essentially all the current flows into the shorted input, and tracing its path with a current tracer will lead to the bad input as shown in FIG. 7-13.

SEQUENTIAL LOGIC

Sequential logic describes logic circuitry that follows a specific order. The devices (combinations of logic gates) can store or delay certain bits, which provides a memory function. Each stage in a sequential logic circuit depends on the results of the previous stage for its own inputs. Such types of circuits include the *flip-flop* (bistable multivibrator), the *one-shot* (monostable multivibrator), the *free-running* or *astable* multivibrator, and the

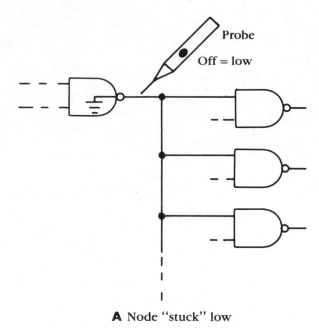

A Node "stuck" low

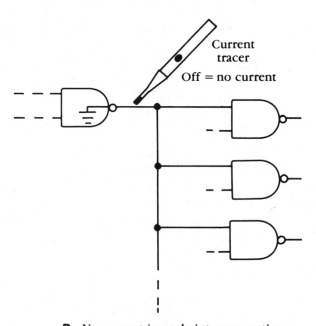

B No current in node interconnections

7-12 Circuit conditions with a shorted output on a driving gate.

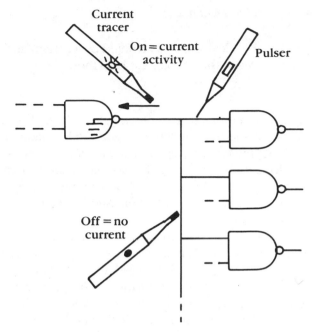

C Tracing current along path into short

7-12 Continued.

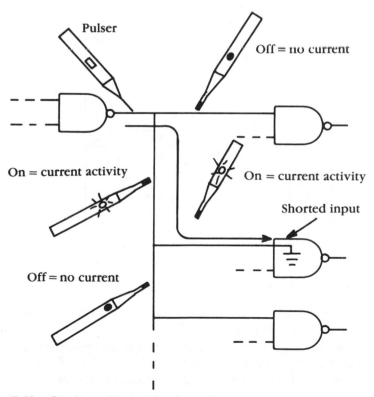

7-13 Circuit conditions with a shorted input on a load gate.

Schmitt trigger (a bistable device that is activated by a certain analog voltage level). The remainder of this chapter explains these devices.

Flip Flops

A *flip-flop* is a bistable circuit made up of logic gates. A bistable circuit can exist in either of two stable states indefinitely and can be made to change its state by means of some external signal. The most important use of this property is a flip-flop's ability to "store" binary information. Because the flip-flop's output remains at either a 0 or a 1 depending on the last input signal, the flip-flop can be said to "remember." Another name for the flip-flop is *bistable multivibrator.*

For purposes of explanation, assume that Q1 in FIG. 7-14 is initially conducting, causing point A to be near ground potential. The base of Q2 will be slightly negative; so Q2 can be considered cut off, resulting in a large positive voltage at point $\overline{A}$. The resultant positive voltage at the base of Q1 further tends to drive Q1 towards saturation, and it can be seen that the stable state for the initial conditions is A = 0V, and $\overline{A}$ = + V.

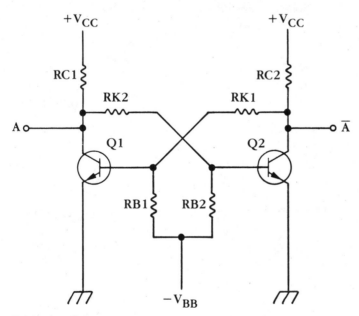

7-14 Flip-flop circuit.

Now assume that for a brief time a ground is placed at point $\overline{A}$. The voltage at the base of Q1 will immediately drop, cutting off Q1 and causing a large positive voltage at point A. A positive voltage applied to the base of Q2 causes saturation in that transistor and point $\overline{A}$ tends to remain at ground potential. Even if the ground is no longer present at point $\overline{A}$ the flip-flop remains in its new stable state, which is A = + V and $\overline{A}$ = 0V. This property (of triggering to either state and remaining there after the triggering signal is removed) accounts for the storage property of the flip-flop.

There are several types of flip-flops and each of these has characteristics that distinguish it from other types. The following descriptions outline the unique characteristics of each type of flip-flop. (A truth table and specific logic symbols are given in each case.) Timing diagrams are included as an aid to understanding the sequential nature of the circuits. As with normal waveforms associated with analog circuits, the timing diagrams are used to present a time-versus-voltage history for the various circuit inputs and outputs.

There are several unique terms used to describe flip-flop operation. There are two complementary outputs from a flip-flop, normally labeled Q and $\overline{Q}$. When one output is in the logic 1 state, the other output is always a logic 0. If the flip-flop changes states, then both Q and $\overline{Q}$ change.

A flip-flop is considered to be *set* when Q = 1 and $\overline{Q}$ = 0. Conversely, the flip-flop is *reset* when Q = 0 and $\overline{Q}$ = 1. Thus, the process of causing the flip-flop to go to the Q = 1, $\overline{Q}$ = 0 state is called *setting* the flip-flop; and causing the flip-flop to go to the Q = 0, $\overline{Q}$ = 1 state is called *resetting* the flip-flop (this action is also referred to as "*clearing*").

NAND Gate SR Flip-Flop   A simple flip-flop can be made from two cross-connected NAND gates as shown in FIG. 7-15. If the two inputs are labeled *S* and *R* (for set and reset), then the flip-flop can be called a set-reset or SR flip-flop. Referring to the figure, it can be seen that if the *S* input goes to a logic 0, the flip-flop will go to its set state (Q = 1) and will remain there until reset. When the *R* input goes to a logic 0, the flip-flop will go to the reset state and stay there until it is set. Thus, an SR flip-flop changes states upon sensing a change in state at the *S* or *R* inputs and stores the results of the change until the opposite input is activated.

There are two special conditions of interest for this flip-flop. First, notice that the condition *S* = 0, *R* = 0 is labeled as being "not allowed." This means that the operation of the circuits that send inputs to *S* and *R* must be restricted such that they never go to a logic 0 at the same time. If this condition were to occur, it can be seen from examination of a basic NAND gate truth table, that both Q and $\overline{Q}$ would go to a logic 1 at the same time. Of course, this is not the defined operation of a flip-flop. Secondly, observe that the conditions *S* = 1, *R* = 1 is the quiescent state for this flip-flop. As long as both inputs remain at a logic 1, no change in the state of the flip-flop is possible.

NOR Gate SR Flip-Flop   The circuit shown in FIG. 7-16 is a flip-flop which is configured by cross-connecting NOR gates in a manner that is very

Logic symbol

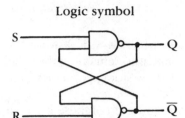

Truth table

S	R	Q
0	0	Not allowed
0	1	1
1	0	0
1	1	No change

Timing diagram

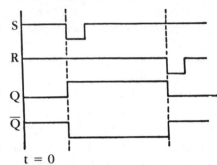

t = 0

7-15 NAND gate SR flip-flop.

Logic symbol

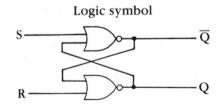

Truth table

S	R	Q
0	0	No change
0	1	0
1	0	1
1	1	Not allowed

Timing diagram

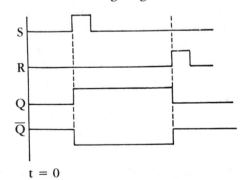

t = 0

7-16 NOR gate SR flip-flop.

similar to the NAND gate version previously described. The main difference between the two is that the activity changes at the S and R inputs are logic 1's instead of logic 0's. Therefore, for the NOR gate configuration, the quiescent state is $S = 0$, $R = 0$, such that no change will occur and the *not allowed* state is $S = 1$, $R = 1$. Since this state is not allowed, the output for this condition is often referred to as undefined or *don't care*. This state becomes important when considering logic minimization techniques.

Clocked Flip-Flops To describe the operation of clocked flip-flops, the concept of a synchronous system is introduced. A synchronous system is controlled by a master oscillator and wave-shaping circuit that produces a set of *clock* pulses. These clock pulses occur at some fixed interval (for example, every 10 microseconds) and all logic state changes are synchronized to occur at the time when the clock pulse occurs. Clocked flip-flops are the means by which this synchronization is maintained.

The truth table for a clocked flip-flop is defined in terms of an output (Q) at the next clock pulse if the state of the output is known at the present clock pulse. Thus, if the output of the flip-flop for clock pulse n is some arbitrary state Q—that is, either one or zero—then at clock pulse $n + 1$ the new output is defined. The new output might be state Q (no change from the previous clock period) or state $\overline{Q}$ (the opposite from what it was); it could be known to be a logic 1, a logic 0, or perhaps an undefined state. As previously noted, an undefined output state merely denotes a *not allowed* combination of inputs.

Clocked SR Flip-Flop Basically, a clocked SR flip-flop operates in the same manner as the NAND and NOR gate versions with the exception that the clock pulse controls the times at which state changes can happen. Referring to the timing diagram of FIG. 7-17, it is seen that the Q and $\overline{Q}$ outputs do not respond directly to the S and R inputs; rather, they await the next clock pulse before changing states. For this flip-flop, if the output is in state Q when clock pulse n occurs, it will remain in that state for clock pulse $n + 1$ provided that inputs S and R are both zero. If either S or R goes to the logic 1 state, then the flip-flop will assume the appropriate state upon occurrence of the next clock pulse. Finally, as with the other SR flip-flops, there is a *not allowed* condition, namely $S = 1$, $R = 1$, for which the output is undefined. This condition is represented by a question mark in the truth table.

T Flip-Flop A commonly required function is performed by the T flip-flop shown in FIG. 7-18. This flip-flop has the ability to change states (toggle) each time a clock pulse occurs. The T flip-flop has one input, T, which controls whether or not the toggling action will occur. When T is set to a logic 1, the flip-flop toggles, and when T is a logic 0, the flip-flop remains

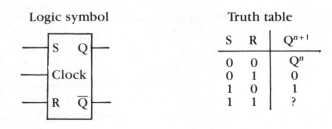

Logic symbol

Truth table

S	R	Q^{n+1}
0	0	Q^n
0	1	0
1	0	1
1	1	?

Timing diagram

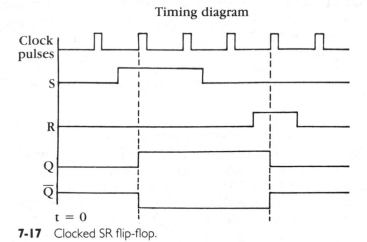

7-17   Clocked SR flip-flop.

in its current state. This ability to toggle with each clock pulse is the basis for digital counters and frequency dividers.

Referring to timing diagram A, it is seen that the result of changing state each time is, in effect, a divide-by-two of the basic clock frequency. This is shown even more graphically in timing diagram B, where a wide clock pulse is used; it can be observed that the output (Q) is indeed a square wave at one-half the frequency of the clock pulse.

Careful examination of the states of the two waveforms, clock vs. outputs, shows that while T is a logic 1, the waveforms go through a binary count sequence. That is, the states are 00, 01, 10, and 11, in that order. Here, the first digit represents the Q output states and the second digit represents the clock pulse states. In other words, the T flip-flop operates as binary counter. If several T flip-flops are connected together in an appropriate manner, any desired count sequence of frequency division can be obtained. Various counter configurations are described in subsequent chapters.

D Flip-Flop   If an inverter is placed at the *R* input to an *SR* flip-flop, as shown in FIG. 7-19, the result is a *D* or delay flip-flop. Since the *S* and *R* inputs are always in opposite states, the flip-flop merely follows the state

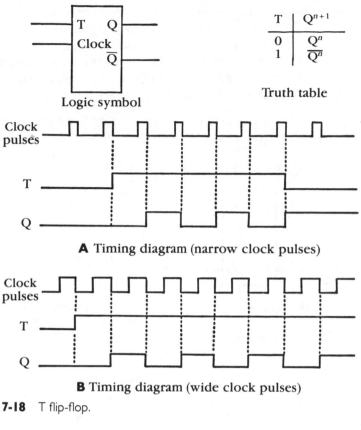

T	Q^{n+1}
0	Q^n
1	$\overline{Q^n}$

Truth table

Logic symbol

A Timing diagram (narrow clock pulses)

B Timing diagram (wide clock pulses)

7-18 T flip-flop.

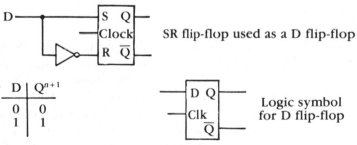

SR flip-flop used as a D flip-flop

D	Q^{n+1}
0	0
1	1

Logic symbol
for D flip-flop

7-19 D flip-flop.

of whatever data is present at the *D* input. More importantly, the result of this action is to delay the data at the input by one clock pulse. In other words, the flip-flop can be considered to have stored the data from the time of the first clock pulse until such time as another clock pulse occurs.

However, *D* flip-flops are not necessarily built from *SR* flip-flops, and a circuit that operates as a D flip-flop may not be capable of performing as an SR flip-flop. Thus, a logic symbol as shown in the figure is used to depict a *D* flip-flop. An important usage of D flip-flops is in the intercon-nection of several flip-flops in series to form *shift registers* or in parallel to form *data storage registers*. In the shift register configuration, delays of many clock periods can be obtained. When used as storage registers, the flip-flops can store data indefinitely, or until required. Shift registers and storage registers have many applications in digital data handling circuits.

JK Flip-Flop   Perhaps the most useful of all the flip-flop configurations is the JK flip-flop, the symbol for which is shown in FIG. 7-20. A comparison

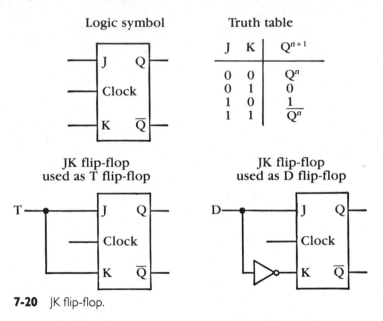

Logic symbol		Truth table		

J	K	Q^{n+1}
0	0	Q^n
0	1	0
1	0	1
1	1	$\overline{Q^n}$

7-20 JK flip-flop.

of the truth tables for the JK flip-flop with that for the clocked SR flip-flop reveals that the difference between the two is that the $J = 1, K = 1$, con-dition is allowed, whereas the $S = 1, R = 1$ condition is not allowed. In the JK flip-flop, the $J = 1, K = 1$ condition causes the flip-flop to toggle. With this one modification, the JK flip-flop is created which can be used for any of the previously described flip-flop functions. For example, it is obvious that the JK flip-flop can be used as an SR flip-flop, since the only difference between the two is a state that is not allowed for SR operation anyway. Further, if the JK inputs are tied together as shown, the truth table reduces to that of the *T* flip-flop previously described, and functionally the two circuits are identical. Finally, if an inverter is added, such that *J* is always the inverse of *K*, the truth table reduces to that of the D flip-flop. Again, the two circuits are functionally identical. JK flip-flops are often used as the basic flip-flop element in a system design due to their ability to be used in these various ways.

MOS Dynamic Storage

A completely different type of storage device from flip-flops is the dynamic storage obtained from the use of metal oxide semiconductor (MOS) transmission gates in conjunction with MOS inverters. All storage with a conventional flip-flop is considered static, in that data stored in a flip-flop can be held indefinitely. The storage obtained through the MOS transmission gates, however, is considered *dynamic,* in that data can only be held for a limited time before it must be refreshed (stored again).

In essence, the MOS transmission gate is a single-pole, single-throw semiconductor switch which can be used to pass current to a capacitor and then be shut off to prevent the capacitor from discharging at a later time. However, the key to the whole storage scheme lies in the fact that an actual capacitor is not required. Happily, the gate terminal of a MOS transistor looks like a capacitor with almost no leakage. Thus, the transmission gate is used to charge up the gate capacitance of another MOS transistor in an inverter stage; then, when the gate closes, the charge is stored in that stage and the stored charge does not leak off.

To illustrate how the transmission gate is employed to charge this capacitance, consider for example the n-MOS transmission gate shown in FIG. 7-21. Assume that the input signal is some positive voltage at the drain terminal and that the clock pulse is initially low. With the clock pulse low, the gate is off and the capacitive load remains uncharged. Now, when the clock goes positive, the gate turns on and the load capacitance rapidly

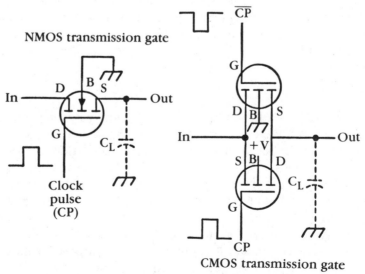

7-21 MOS transmission gates.

charges. At a still later time, when the clock goes low, the gate opens up again and the charge is now stored by the load capacitance. A (complementary MOS) CMOS transmission gate is also shown for reference; however, its operation is quite similar to that of the n-channel device.

It is apparent that a low-speed limit and a high-speed limit are both imposed on this circuit. At the low-frequency end, it must be remembered that there is some small amount of leakage that will occur. If the clock frequency is not sufficient to read out data faster than the capacitive decay rate, the proper logic levels will no longer be present when an attempt is made to read the data. A high-frequency limit exists because a finite amount of time is required to charge the gate capacitance. If the clock frequency is too fast, the capacitance will not become charged sufficiently, and again the proper logic levels will not be present.

If the transmission gate of FIG. 7-21 were simply used with an inverter and a series connection made of a group of these circuits, there would be no storage at all. This is because at the next clock period the load capacitance would be discharged by the previous stage at the same time it was being read into the following stage.

This problem is solved by using a two-phase clocking system similar to that shown in FIG. 7-22. Here, input data is determined such that it changes states at the phase-two clock leading edge. Thus, when the phase-one clock occurs, the input data is settled to its desired state and the capacitance of the first inverter can charge up. After the phase-one clock goes low, the first inverter has stored the signal until the phase-two clock pulse occurs. At this time, the first inverter charges the second inverter capacitance through the transmission gate, and the charge is now stored at the second inverter. This series performance can be repeated for as many of these two-phase stages as desired, each stage providing a delay equal to one clock period.

One-Shot

As previously described, the flip-flop has two stable states and can remain in either one indefinitely. The one-shot, however, has only one stable state and another state called a quasi-stable state. In operation, the one-shot remains in its stable state until a triggering signal is received. Upon receipt of the triggering signal, the one-shot changes to the quasi-stable state for a fixed period of time, then returns by itself to the stable state again. Since the circuit always returns to its single stable state, it is called a one-shot or single-shot. Another name for this device is a *monostable multivibrator.*

The circuit shown in FIG. 7-23 will be used for purposes of illustrating the operation of a one-shot. Starting with the stable state of A = 0V and $\overline{A}$ = + V, circuit operation will be analyzed. With A = 0V, the base of Q1

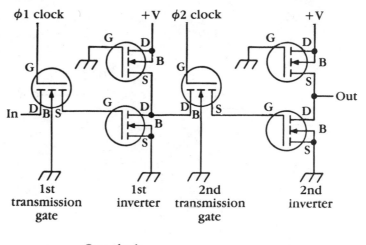

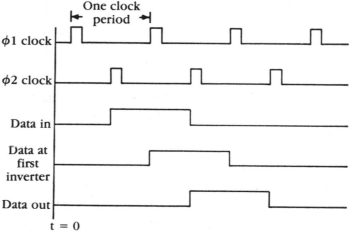

7-22   Single stage of MOS dynamic storage.

is held slightly negative, thus keeping Q1 cut off. Point $\overline{A}$ at $+$ V keeps both sides of capacitor C at the same voltage potential and the base of Q2 is held positive, causing Q2 to be in saturation.

Now assume that a ground is placed at point $\overline{A}$ very briefly. The ground will immediately cause capacitor C to start charging through R_t, forcing the base of Q2 to go negative. This action results in a change of state to the quasi-stable condition where A $=$ $+$ V and $\overline{A}$ $=$ 0V. However, as capacitor C continues to charge, the base voltage at Q2 will eventually rise sufficiently to cause Q2 to once again resume conduction. The length of time for this to happen is based on the time constant for R and C. The time in seconds is equal to the value of R (in megohms) times the value of C (in microfarads). When the conclusion point is reached, the one-shot

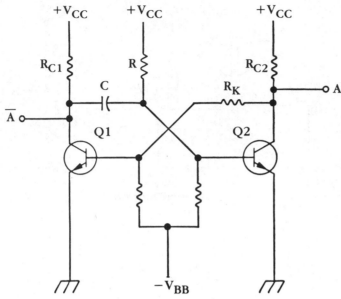

7-23   One-shot circuit.

will return to its stable state. From the above, it can be seen that a one-shot produces an output starting with a triggering signal and ending at a time based on the RC time constant of the circuit.

Free-Running Multivibrator

The *free-running* or *astable* multivibrator does not have a stable state; instead, it continually switches between one state and the other. The time that it stays in each state is dependent on the RC time constants of the individual resistors and capacitors used to make up the multivibrator. The most common use for a free-running multivibrator is as an oscillator that generates rectangular timing pulses for use throughout a digital unit.

A basic free-running multivibrator circuit is shown in FIG. 7-24. Assume that initially Q1 is switched on and Q2 has just switched off. Capacitor C1 will immediately present a large positive potential at the base of Q1, holding it in saturation, while capacitor C2 couples essentially 0V to the base of Q2. However, as time goes on, the two capacitors charge, the base of Q1 will tend to become more positive, and the base of Q2 will simultaneously become less positive. Hence, at some point, Q1 will start to conduct less while Q2 begins conducting more.

This effect is regenerative because the changes in conduction are reflected at the two collectors, whose voltages are in turn coupled back to the bases of the opposite transistors. As a result, the circuit will switch very rapidly to the state where Q1 is cut off and Q2 is in saturation. At this point,

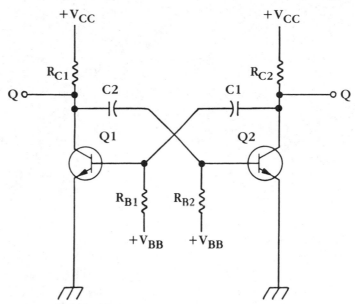

7-24 Free-running multivibrator.

the slow charging procedure is again initiated, this time in the opposite direction. As the potential at the two bases begin to change significantly, the regenerative action again causes a very rapid switching of states, back to the initial conditions. The above two cycles repeat indefinitely, with the timing of the cycles based upon the respective time constants, $R_{B1}C1$ and $R_{B2}C2$.

Schmitt Trigger

A *Schmitt trigger* is a bistable digital device whose state is dependent upon the amplitude of an analog input voltage. When the input voltage is below a predetermined threshold, the output is a logic 0. If the input rises above a second threshold, the output switches to a logic 1. The Schmitt trigger circuit is often used as a threshold detector to determine when some unknown input voltage has crossed the given threshold voltage. Another common use of a Schmitt trigger is to make square waves out of sine waves.

A Schmitt trigger circuit is shown in FIG. 7-25. To understand operation of the circuit, assume that initially the input to transistor Q1 is some negative voltage such that Q1 is cut off. Then a voltage divider, consisting of resistors R_{C1}, R_K, and R_{B2}, is formed as shown. This voltage divider establishes a positive threshold voltage at the base of Q2, dependent upon the resistive values in the divider. This positive threshold voltage will also be present at the common emitter resistor R_E. However, since the two transis-

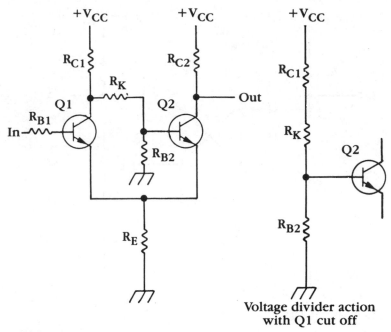

7-25   Schmitt trigger.

tors share a common emitter resistor, the effect is one of regenerative feedback, where the positive emitter voltage serves to keep transistor Q1 in its cutoff state. Also, since the two transistors have the same emitter voltage, it should be apparent that the only way to make the circuit change state is to place a more positive voltage at the base of Q1 than the voltage that is already at the base of Q2, due to the voltage-divider action. Hence, the threshold is, as previously stated, entirely dependent on the resistance values in the voltage divider.

As soon as the input voltage rises above this threshold, transistor Q1 will forward bias and the voltage at the base of Q2 will immediately drop. Through the regenerative feedback of resistor R_E, a new stable state will be achieved with Q1 on and Q2 off.

Now, the voltage divider still exists; however, the voltage at the junction of R_{C1} and R_K is somewhat lower than before due to Q1 conducting. What this means is that even if the input voltage drops back down to the previous threshold level, the base of Q1 will still be more positive than the base of Q2 and the Schmitt trigger will not change states. What must occur is for the input to drop even lower until the base of Q1 is indeed more negative than the base of Q2.

At this time, Q1 will cut off, causing the previous voltage-divider action; transistor Q2 will turn on. As before, the regenerative feedback of R_E

assures that the Schmitt trigger remains in this stable state. Note that if the input voltage is not changed, the base potential is not now sufficient to cause a state change again. Hence, the voltage difference between the two threshold voltages, called the *hysteresis voltage,* assures that the circuit will not oscillate at some particular voltage.

Thus, it is seen that a Schmitt trigger changes states at two separate (but related) threshold voltages. The more positive threshold must be crossed before the Schmitt trigger will switch from the logic 0 state to the logic 1 state. Similarly, the more negative threshold must be crossed to cause the Schmitt trigger to switch from the logic 1 state to the logic 0 state.

If power were applied with an input voltage between the two threshold values, the circuit would assume one of its stable states (based upon the input voltage) and then would not switch until one of the thresholds was crossed.

Summary

There are two basic types of logic operations: combinational and sequential. In combinational logic, inputs are assumed to be constant at any given time, and the output is a function only of the inputs. Combinational logic circuits use truth tables, and they are simplified by the use of Boolean algebra or Karnaugh maps. One of the objectives when using combinational circuits is to reduce the number of gates and the number of levels involved, thus reducing cost and propagation delays. Troubleshooting of combinational logic circuits is often accomplished by using a logic pulser and a logic probe, to determine whether specific inputs or outputs are open or shorted.

Sequential logic circuits refer to those circuits in which inputs and outputs occur in a specific order, and outputs depend not only on inputs but on outputs of other stages and previous states as well.

Three primary examples of sequential logic circuits are the flip-flop or bistable multivibrator, the monostable multivibrator, and the astable multivibrator. The Schmitt trigger is a special case of a bistable device.

Questions

1. Explain the difference between combinational and sequential logic and give an example of each.

2. Identify two ways in which to reduce combinational networks.

3. What is gate minimization, and why is it important when dealing with combinational networks?

4. What are the two main tools used to troubleshoot gate networks? What does each do?

5. What is a flip-flop? What is the difference between a clocked and an unclocked flip-flop?

6. What is the primary special characteristic of the JK flip-flop?

7. Give the primary characteristics of each of the following: one-shot (monostable multivibrator) and free-running (astable) multivibrator.

8. What is a Schmitt trigger?

***

Problems

1. For the circuit in FIG. 7-26,   **a.** write the Boolean expression
   b. write the truth table

2. Repeat Problem 1 for the circuit in FIG. 7-27.

3. Draw a logic circuit for the expression $X = A + B (C + AD)$.

4. Draw a logic circuit for the expression $X = AB + C (A + BD)$.

5. Given the circuit in Problem 3,   **a.** determine the number of logic levels   **b.** simplify using Boolean algebra   **c.** simplify using a Karnaugh map   **d.** draw the simplified diagram

6. Given the circuit in Problem 4,   **a.** determine the number of logic levels   **b.** simplify using Boolean algebra   **c.** simplify using a Karnaugh map   **d.** draw the simplified diagram

7. For the circuit in FIG. 7-26, output X is always high. Where might the trouble be? How could this be tested?

8. For the circuit in FIG. 7-27, input C appears to have no effect on the output, independent of inputs A and B. Where might the trouble be? How could this be tested?

9. For FIG. 7-28, input 1 is the S input and input 2 is the R input for a clocked SR flip-flop. Sketch the output.

10. For FIG. 7-29, input 1 is the J input and input 2 is the K input for a JK flip-flop. Sketch the output.

11. If FIG. 7-30 shows a D-type flip-flop, sketch the output. (Ignore input 2.)

12. If FIG. 7-30 shows a T-type flip-flop, sketch the output.

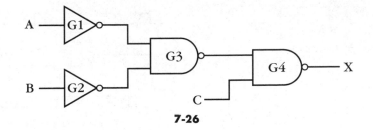

7-26

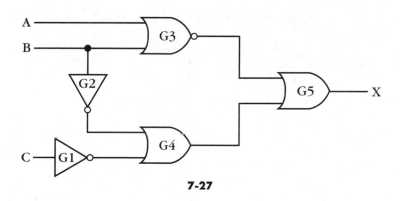

7-27

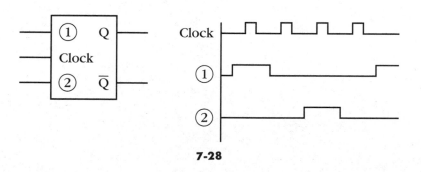

7-28

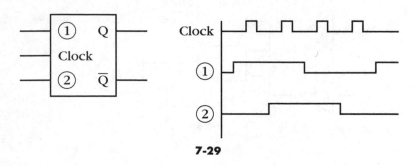

7-29

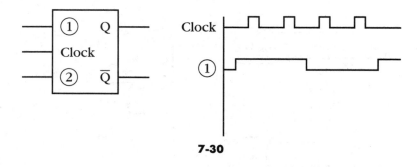

7-30

TYPICAL LOGIC NETWORKS

AFTER YOU COMPLETE this chapter, you will be able to:

☐ Understand how shift registers are used for different functions
☐ Describe counters of various types
☐ Discuss decoders, encoders, and code converters

There are many different types of logic networks, and every digital system has numerous networks that are unique to that system. There are, however, a few logic networks that are encountered again and again. These networks may vary in minor detail from system to system, but their primary functions are always the same. This chapter introduces a few of the typical logic networks found in almost any digital system. Among the typical logic networks described are registers, counters, adders, and code converters.

REGISTERS

One of the most important uses for flip-flops in logic networks is to form *registers*. Registers are the means by which digital data is stored for use at appropriate times by other logic networks. The data may be stored for a very short time (one clock period) to permit its use in some logic operation; or it may be stored for a longer period (several clock periods, where the total elapsed time may be measured in milliseconds) to provide an output after a certain sequence of operations; or it may be stored indefinitely to represent such things as equipment status. Whatever the purpose or time duration of the storage, the same types of registers are used to perform this storage.

Perhaps the simplest type of register that can be configured is the *serial shift* register. This register simply shifts the data bits contained in the individual flip-flops one stage to the right each time a clock pulse occurs. A logic diagram of the basic shift register configuration is shown in FIG. 8-1. This register is made up with JK flip-flops, but there is no reason that

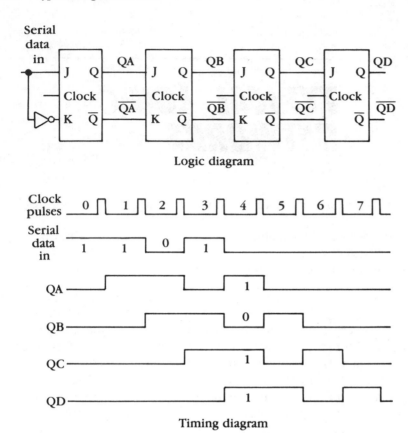

Logic diagram

Timing diagram

8-1   Basic shift register configuration.

D flip-flops or SR flip-flops could not perform the identical functions. To describe the operation of this circuit, consider the timing diagram shown directly below the logic diagram. A serial data stream is presented at the shift register data input. The data stream in this example consists of the bits 1101, which occur sequentially, the most significant bit first. Assuming that the register is initially in the 0000 state, a logic 1 is read into flip-flop A by the first clock pulse, causing the state of the register to become 0001. At the second clock pulse, the data is shifted right one bit, so that the contents of flip-flop A are now contained in flip-flop B. At the same time, the next serial data bit is read into flip-flop A. Thus, after clock pulse two, the state of the register is 0011.

The third clock pulse produces a similar right shift, changing the register state to 0110. The fourth clock pulse reads in the final data bit, while again shifting data in the other flip-flops one place to the right, and the state of the register becomes 1101. At this point, the serial data is now fully contained in the register and a number of options are available:

1. The individual flip-flop outputs may be read out in parallel during the fourth clock period. In this manner, the data can be transferred to other logic circuits.

2. The clocks to the register can be stopped. Here, the register is used as a temporary storage register, storing the data until needed.

3. The output of flip-flop D can be used as a delayed serial data stream. In this case, the register acts as a four-clock-period delay line.

A more general-purpose register configuration is shown in FIG. 8-2. This register features a direct set and reset capability which is frequently found in integrated circuit flip-flops and registers. Basically, the direct set and reset inputs override all other inputs to enter data into the flip-flops. Thus, for example, no matter what actions are occurring at the JK inputs, if a pulse is applied to the clear input, the register will assume the 0000 state.

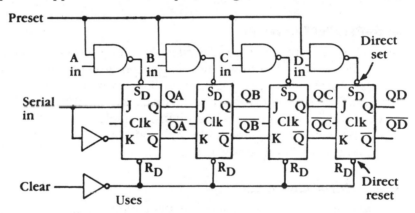

Uses
- Delay line
- Storage register
- Serial-to-parallel converter
- Parallel-to-serial converter

8-2 Multipurpose register configuration

Similarly, once having been cleared, any desired data bits can be parallel entered into the register through the A, B, C, and D inputs, if a logic 1 is also present at the preset input. There is no requirement that a clock pulse be present. Thus, the direct set and reset inputs allow parallel data entry into the register, with a resultant increased flexibility for various uses. Consider, for example, some typical uses of this register.

Delay Line

If operated as a serial shift register, the register acts as a delay line, providing a delay equal to the number of flip-flops in the register. One frequency application of a digital delay line is to align several different sets of digital data so that their bit positions agree. After alignment, arithmetic or logic operations can be performed on the data.

Storage Register

Data can be entered in serial form or through the parallel inputs for storage. As long as the flip-flops receive no clock inputs after the data is in the register, the flip-flops will store the data for as long as desired. Alternatively, the output of flip-flop D can be connected back to the register input to provide a recirculating register that shifts the data in a continuous loop. Care must be taken to keep track of the number of shift clocks that have been received so that the proper location of data can be determined.

Serial-to-Parallel Converter

Serial data is shifted into the register until it is properly aligned, then read out in parallel format. A common reason for converting from serial to parallel data format is so that logic operations can be performed in a shorter period of time. Since each bit of serial data requires one clock period to become available, it can be seen that a simple function like adding two 10-bit numbers requires 10 clock periods. In parallel format, this operation requires only one clock period.

Parallel-to-Serial Converter

By using the direct set and resets inputs to this circuit, parallel data can be entered into the register. If clocks are now applied to the register, the output of flip-flop D is a serial data stream representing the parallel input data. One reason for converting to serial data format is the transmission of digital data. In parallel format, a 10-bit number requires 10 line driver circuits and 10 wires in a cable. In serial format, only one driver and one wire is required.

COUNTERS

In addition to registers, almost any digital device will have one or more counters. Counters, like registers, are also made up of flip-flops and associated gating. In general, the maximum number of states that a counter can have is 2^n, where n is the number of flip-flops in the counter. For example, a counter with two flip-flops has $2^2 = 4$ states. A three-stage counter could have a maximum of $2^3 = 8$ states. The actual number of counts defined by a particular design is called its *modulo* count. For example, a *modulo seven* counter has 7 counts in its count sequence.

There are two basic types of counter construction: ripple and synchronous. A ripple counter is the simpler to build and uses the output from each preceding stage as the clock for the next stage. However, if there are quite a few stages to the counter, an excessive amount of delay time may be encountered between the time the first stage changes and the time when the last stage performs its transition.

To overcome this problem, synchronous counters are used when speed is an important factor. A synchronous count is obtained by using additional logic gates to ensure that all stages of the counter change states at the same time.

Ripple Counter

An example of a typical ripple counter is, shown in FIG. 8-3. This counter is *modulo eight* because it has 8 counts and it is binary because it counts in a pure binary sequence. It is a ripple counter because the clock for each stage is the output from the previous stage.

There are several points that should be kept in mind when analyzing this circuit: First, remember that flip-flops change states at either the leading or the trailing edge of the clock pulse. The flip-flops shown here change states on the negative-going clock pulse edge. Secondly, a key to this circuit's operation is the ability of a JK flip-flop to toggle if its *J* and *K* inputs are both a logic 1. Since positive-true logic is being used throughout this book, $+V$ represents the logic 1 condition. Hence, each stage toggles, and the rate of toggling is always at half the clock frequency. Thus, flip-flop A divides by two, flip-flop B divides by four, and flip-flop C divides by eight. From the timing diagram, each successive division occurs on the trailing edge of the clock pulse. Note also that all the counts in the state table can be read from the timing diagram, if desired. The state table merely provides a more convenient and compact notation for the information, and usually the minute details of the timing diagram need not be considered to understand the operation of the counter.

Synchronous Counter

A synchronous counter is shown in FIG. 8-4. This counter is *modulo sixteen* because it has 16 counts and is binary because it counts in a binary sequence. The counter is synchronous because all four stages change states at the same time at the trailing edge of the clock pulse. Thus, there is no ripple delay time as was the case in the ripple counter.

The operation of this counter is quite different in that, instead of tog-

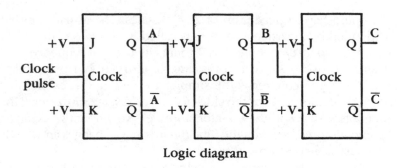

Logic diagram

Clock Pulse	C	B	A
0	0	0	0
1	0	0	1
2	0	1	0
3	0	1	1
4	1	0	0
5	1	0	1
6	1	1	0
7	1	1	1

State table

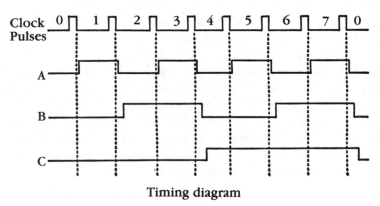

Timing diagram

8-3   Modulo-eight binary ripple counter.

gling the flip-flop on a trailing edge of some signal generated by a previous flip-flop, a desired state is detected and the flip-flop is caused to toggle on the next clock pulse after the detected state occurs. Since the previous stage also has its transitions on the trailing edge of the clock pulse, the two stages change states at the same time, or *synchronously.*

Logic diagram

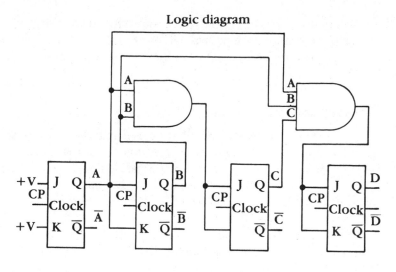

Truth table

Clock Pulse (CP)	D	C	B	A
0	0	0	0	0
1	0	0	0	1
2	0	0	1	0
3	0	0	1	1
4	0	1	0	0
5	0	1	0	1
6	0	1	1	0
7	0	1	1	1
8	1	0	0	0
9	1	0	0	1
10	1	0	1	0
11	1	0	1	1
12	1	1	0	0
13	1	1	0	1
14	1	1	1	0
15	1	1	1	1

8-4 Modulo-sixteen binary synchronous counter.

Examining the state table, each clock pulse is seen to cause flip-flop A to toggle. Flip-flop B toggles one clock period after flip-flop A has been set to a logic 1. Similarly, flip-flop C is caused to toggle one clock pulse after flip-flops A and B are both set to the logic 1 level. Flip-flop D operates in the same way, requiring that a change of state take place one clock pulse after detecting that flip-flops A, B, and C are all set to the logic 1 state. Each of these requirements can be readily seen by noting the count sequence that the state table must follow if a binary count is to be obtained.

Interestingly, even though the techniques for binary and ripple counters are quite different, their state tables are identical. As previously mentioned, the main difference between the two is their speed of operation.

BCD (Decade) Counter

The counters previously described all used the maximum number of counts available, that is, 2^n states. However, counters can readily be made to count any desired number of counts by providing the proper gating to skip some counts that would normally occur. One such counter commonly used is the BCD or decade counter. A decade counter is a *modulo ten* counter (because it has 10 counts).

A typical decade counter that counts in 8421 BCD is shown in FIG. 8-5. This counter is very similar to the synchronous modulo sixteen counter

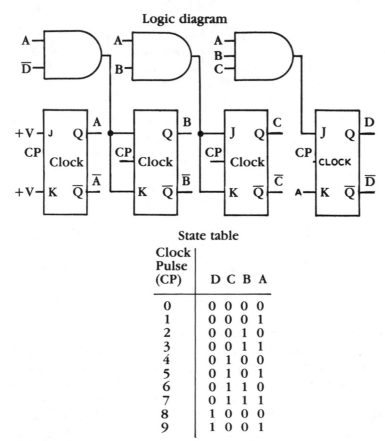

Logic diagram

State table

Clock Pulse (CP)	D	C	B	A
0	0	0	0	0
1	0	0	0	1
2	0	0	1	0
3	0	0	1	1
4	0	1	0	0
5	0	1	0	1
6	0	1	1	0
7	0	1	1	1
8	1	0	0	0
9	1	0	0	1

8-5 8421 BCD synchronous counter.

previously described except that the logic gating has been modified in such a way that the last six counts are skipped.

To see the skip action, consider the natural binary sequence presented by the modulo sixteen counter. As the counter reaches the ninth count (1001), the next state is the tenth count (1010); but instead, it is desired that the next count in sequence be zero (0000). The states of flip-flops A and C will be correct and need not be changed. Flip-flop B, however, will toggle to the 1 state unless this action is inhibited. Thus, the toggle term for flip-flop B is modified from $J = $ A, $K = $ A to $J = $ A$\overline{D}$, $K = $ A$\overline{D}$. here, $\overline{D}$ stops flip-flop B from toggling after the count reaches nine.

Flip-flop D presents a different problem in that it is already set and must be reset after the ninth count. Since the only time that flip-flop D needs to be set is after the seventh count, and the only time it needs to be reset is two counts later, separate set and reset terms are called for. The set term is count seven, the same logic gating which was previously used. The reset, however, can occur each time flip-flop A goes set and, in particular, it is required at the clock pulse immediately following the ninth count. By making these simple modifications, the desired 10 counts are obtained in the 8421 code and the remaining 6 counts are skipped.

Johnson Counter

A very simple counter that uses no logic gates is the Johnson counter. A Johnson counter is basically a shift register with a very unique feedback to its serial input. The feedback is configured such that whatever the state of the output stage, the complement of that state is applied to the serial input at the next clock pulse.

A modulo 10 Johnson counter is shown in FIG. 8-6. Note from the state table that logic 1 levels are shifted into the register for five counts, then logic 0 inputs are shifted for the next five counts, etc. An important feature of the Johnson counter is that the cycle length is $2n$, instead of 2^n. Hence, for the example shown there are five stages, so that $n = 5$, and the resultant cycle length is $2n = 10$. Because there are $2n$ counts and 2^n possible states which n flip-flops can assume, the counter can be in any one of these $2n$ states when power is applied.

If the state that occurs is not one of those in the normal count sequence, the counter will not count properly and it may never correct itself to the proper sequence. One method for ensuring that the counter will operate properly is with a clear input. Upon application of power, a logic 1 is momentarily applied to the clear line. This sets the register to all zeros; as seen from the state table, this is one of the allowed counts. Upon removal of the logic 1 from the clear input, the counter will then count correctly.

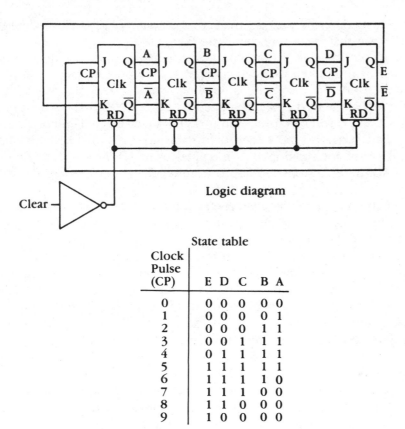

Logic diagram

Clear

State table

Clock Pulse (CP)	E	D	C	B	A
0	0	0	0	0	0
1	0	0	0	0	1
2	0	0	0	1	1
3	0	0	1	1	1
4	0	1	1	1	1
5	1	1	1	1	1
6	1	1	1	1	0
7	1	1	1	0	0
8	1	1	0	0	0
9	1	0	0	0	0

8-6   Johnson decade counter.

Ring Counter

A ring counter is another version of a shift register configuration which uses significantly less than the maximum number of states that can occur. The counter is simply a recirculating shift register into which a single 1 has been entered. Thus, for n flip-flops, the counter has n states. A typical ring counter is shown in FIG. 8-7. From the state table, it is apparent that each count has a single 1 and four 0's. The 1 shifts from stage to stage in an endless loop.

As with the Johnson counter, a ring counter must be initialized before it will operate properly. A method which can be used in the counter shown consists of presenting a clear pulse followed by a preset pulse to the input lines. When the clear pulse occurs, the counter will go to the 00000 state. The preset inputs are wired so that when the preset pulse occurs, the counter assumes the 00001 state. From this point on, the counter operates in its normal sequence for each clock pulse received. If

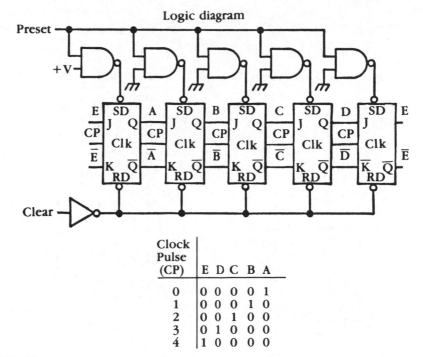

Logic diagram

Clock Pulse (CP)	E D C B A
0	0 0 0 0 1
1	0 0 0 1 0
2	0 0 1 0 0
3	0 1 0 0 0
4	1 0 0 0 0

8-7 Ring counter.

at any time it is desired to restart the counter at its initial count, the procedure must be repeated.

DECODERS

Many applications for the counters just described use the flip-flop outputs directly to perform some type of control function or to form a binary or BCD number. However, often a requirement exists to know when a particular count of the counter occurs and to use this count to start or terminate a particular sequence of events. The method by which the occurrence of a specific count is known is by decoding the desired counter state. Each decode is simply an AND function with its inputs equal to the state being decoded.

Single Decode Gate

An example of a typical decode gate is shown in FIG. 8-8. This particular gate represents a decode of count nine of the modulo 16 binary counter shown in FIG. 8-4. Since a logic 1 is desired for only count nine, the Karnaugh map has only one cell with a logic 1. Also, assuming that the modulo sixteen counter is running continuously, the timing diagram shows the

Truth table

Logic symbol

A
$\overline{B}$
$\overline{C}$
D

f9 = Count nine

D C B A	Count Nine (f9)
0 0 0 0	0
0 0 0 1	0
0 0 1 0	0
0 0 1 1	0
0 1 0 0	0
0 1 0 1	0
0 1 1 0	0
0 1 1 1	0
1 0 0 0	0
1 0 0 1	1
1 0 1 0	0
1 0 1 1	0
1 1 0 0	0
1 1 0 1	0
1 1 1 0	0
1 1 1 1	0

f9 BA   **Karnaugh map**

DC \ BA	00	01	11	10
00	0	0	0	0
01	0	0	0	0
11	0	0	0	0
10	0	1	0	0

$f9 = A\,\overline{B}\,\overline{C}\,D$
Logic equation

Timing diagram

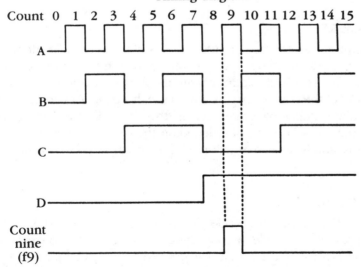

Count 0 1 2 3 4 5 6 7 8 9 10 11 12 13 14 15

A

B

C

D

Count nine (f9)

8-8 Basic decode gate.

shows the resultant decode gate ouput for count nine. Obviously, the logic equation can be written very simply from either the truth table or from the Karnaugh map. It is also evident that any other count can be decoded in a similar manner.

Don't-Care Conditions

An important criterion when determining the requirement for a decode gate is whether or not unused states of the counter exist. Consider, for example, the truth table and associated Karnaugh map for an 8421 BCD count sequence, as shown in FIG. 8-9. It is apparent that only the first 10 out of 16 possible states are used. The remaining states can therefore be considered unused states. If indeed these states never occur, then surely it does not matter if a decode gate is formed such that it will produce an output for one or more of the nonexistent states. Thus, the outputs for the six unused states in this example are called *don't care* outputs, and they are designated by X's in the truth table and Karnaugh map. The beauty of don't care conditions is that when loops are drawn in the Karnaugh map, as many X's as desired can be included in the loop. It has already been shown that the larger the loop, the fewer gates and gate inputs required to implement the function. Including selected don't care conditions in the Karnaugh map loop is equivalent to defining that if the unused states should occur (which they won't), then a logic 1 output will be obtained from the decode gate.

Referring to the Karnaugh map, each f in the map represents a desired output of logic 1 for that count alone. Thus, f9 means that a logic 1 output is desired when count nine occurs. Note that for counts 0 and 1, a normal four-input decode is required. However, counts 2, 3, 4, 5, 6, and 7 all include X's to form a two-variable loop; thus, gates can be formed to decode these counts, which each require one less input.

Finally, it is observed that for counts 8 and 9, a loop of four cells can be made. Thus, these two decades require only two inputs. From this, we see that it is important to recognize don't care conditions and to plot these conditions in the Karnaugh map whenever decode functions are analyzed.

Decodes for Johnson and Ring Counters

Both the Johnson counter and the ring counter typically have many more states than their modulo counts. Since it has just been shown that unused states result in don't care conditions, it might be expected that the very large number of don't cares would simplify decoding considerably. This is exactly the case. Decodes for these two counter configurations are shown in FIG. 8-10.

Truth table

D	C	B	A	Output Decode
0	0	0	0	f0
0	0	0	1	f1
0	0	1	0	f2
0	0	1	1	f3
0	1	0	0	f4
0	1	0	1	f5
0	1	1	0	f6
0	1	1	1	f7
1	0	0	0	f8
1	0	0	1	f9
1	0	1	0	X
1	0	1	1	X
1	1	0	0	X
1	1	0	1	X
1	1	1	0	X
1	1	1	1	X

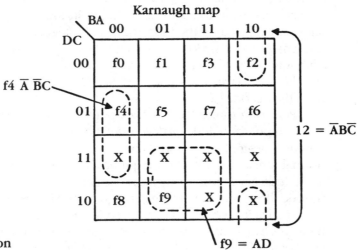

Karnaugh map

Logic equation

$f0 = \overline{ABCD}$
$f1 = A\overline{BCD}$
$f2 = \overline{AB}\overline{C}$
$f3 = AB\overline{C}$
$f4 = \overline{A}BC$
$f5 = A\overline{B}C$
$f6 = \overline{A}BC$
$f7 = ABC$
$f8 = \overline{A}D$
$f9 = AD$

8-9 Decodes for 8421 BCD counter.

Clock Pulse	E D C B A		
0	0 0 0 0 0	f0 = $\overline{AE}$	
1	0 0 0 0 1	f1 = $A\overline{B}$	
2	0 0 0 1 1	f2 = $B\overline{C}$	
3	0 0 1 1 1	f3 = $C\overline{D}$	Johnson counter decodes
4	0 1 1 1 1	f4 = $D\overline{E}$	
5	1 1 1 1 1	f5 = AE	
6	1 1 1 1 0	f6 = $\overline{A}B$	
7	1 1 1 0 0	f7 = $\overline{B}C$	
8	1 1 0 0 0	f8 = $\overline{C}D$	
9	1 0 0 0 0	f9 = $\overline{D}E$	

Clock Pulse	E D C B A		
0	0 0 0 0 1	f0 = A	Ring counter decodes
1	0 0 0 1 0	f1 = B	
2	0 0 1 0 0	f2 = C	
3	0 1 0 0 0	f3 = D	
4	1 0 0 0 0	f4 = E	

8-10 Decoding the Johnson ring counters.

For the Johnson counter it can be seen without even looking at a Karnaugh map that no matter what the Johnson cycle length, it is always possible to decode a particular count with a two-input AND gate. The two inputs to the gate are always such that the decode follows the 1 to 0 or 0 to 1 changeover. The exceptions to this are the all-0 state and the all-1 state. Here, the two end stages of the counter always form the decode.

The decode for a ring counter is even simpler. In fact, it isn't really a decode at all: Since there is always a single 1 in the counter, the stage which has the 1 in it represents the state of the counter. Thus, no gating is required to decode a ring counter. All of the above description can be derived from looking at Karnaugh maps and drawing loops, but in simple cases such as this, intuitive commons sense is the simpler approach.

ENCODERS

An *encoder* is a combinational logic circuit that essentially performs a "reverse" decoder function. An encoder accepts a digit on its inputs, such as a decimal or octal digit, and converts it to a coded output, such as binary or BCD. Encoders can also be devised to encode various symbols and alphabetic characters.

Decimal-to-BCD Encoder

An encoder that converts decimal to BCD has 10 inputs—one for each decimal digit—and four outputs corresponding to the BCD code, as shown in FIG. 8-11. The BCD code is shown in TABLE 8-1, which shows the

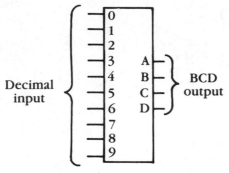

8-11   Decimal-to-BCD encoder block diagram.

Table 8-1.   Decimal and BCD Equivalents.

Decimal Digit	BCD Code D C B A
0	0 0 0 0
1	0 0 0 1
2	0 0 1 0
3	0 0 1 1
4	0 1 0 0
5	0 1 0 1
6	0 1 1 0
7	0 1 1 1
8	1 0 0 0
9	1 0 0 1

relationship between each BCD digit and its corresponding decimal digit. For example, the most significant bit of the BCD code, D, is a 1 (high) for decimal digits 8 and 9. The expression for bit D in terms of the decimal digits can therefore be written as:

$$D = 8 + 9$$

Bit C is a 1 for decimal digits 4, 5, 6, or 7 and can be expressed as follows:

$$C = 4 + 5 + 6 + 7$$

Bit B is a 1 for decimal digits 2, 3, 6, or 7 and can be expressed as:

$$B = 2 + 3 + 6 + 7$$

Finally, A is a 1 for digits 1, 3, 5, 7, or 9, so the expression is:

$$A = 1 + 3 + 5 + 7 + 9$$

Implementing the logic circuitry required to encode each decimal digit to its binary equivalent is simply a matter of ORing the appropriate decimal digit lines to form each output.

Refer to FIG. 8-12. When a high appears on one of the decimal digit

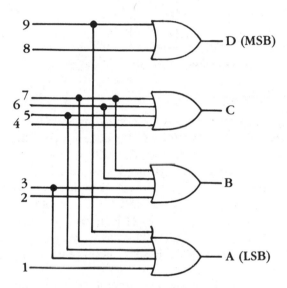

8-12 Logic diagram for the decimal-to-BCD encoder.

lines, the appropriate levels occur on the four BCD output lines. For instance, if input line 9 is high (and all other outputs are low), this produces a high on outputs A and D and lows on outputs B and C, which is the BCD code for 9 (1001).

Octal-to-Binary Encoder

This commonly used type of encoder is often referred to as an eight-line-to-three-line encoder. The same logic can be used in it as the decimal-to-BCD encoder, except that inputs 8 and 9 are omitted because there are only eight octal digits (0 through 7). Also, only three binary digits are re-quired to represent the eight octal digits, so this type of encoder has only three output lines. A block diagram for this type of encoder is shown in FIG. 8-13, and a logic diagram is in FIG. 8-14.

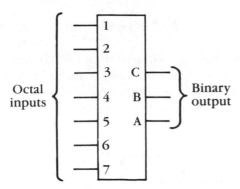

8-13   Octal-to-binary encoder block diagram.

CODE CONVERTERS

Throughout this book, quite a few different binary codes have been de-scribed, each having unique advantages. Because any particular digital sys-tem may use one or more of these codes, a frequently encountered re-quirement is for a logic network that can convert from one code format to another. A circuit that performs this function is called a *code converter*.

Several code converters have already been described. For example, the collection of 10 decodes for the 8421 BCD counter, shown in FIG. 8-9, is often referred to as a BCD-to-decimal code converter. The inputs to the code converter are the four BCD bits, and its outputs are the ten decimal digits.

Another code converter that was previously described is the serial gray-to-binary code converter shown in FIG. 6-9. Many other code convert-ers are possible, several of which are described in the following para-graphs.

Decimal-to-BCD Converter

One reason for converting a BCD number to decimal might be to light some decimal display indicating which of 10 possible actions had oc-curred in a circuit. The decimal readout can then be interpreted by some-one not familiar with binary number codes. Similarly, occasion arises to enter decimal data via switches into the digital circuits. Ten bits of data are very wasteful when the same information can be represented by four bits

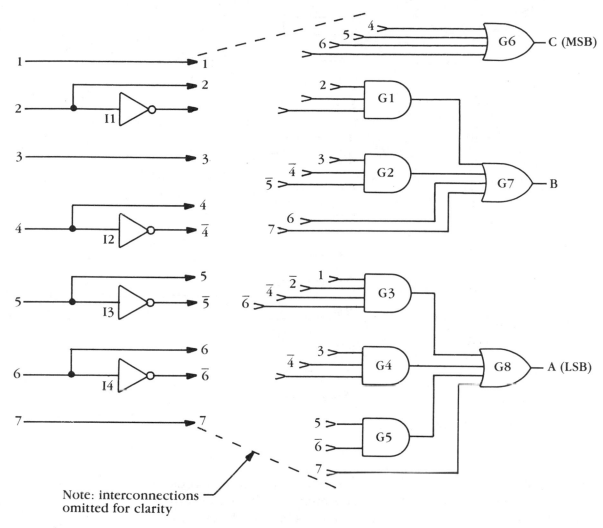

8-14 Logic diagram for the octal-to-binary encoder.

in a BCD code. For this, a decimal-to-BCD code converter similar to that shown in FIG. 8-15 is used.

This circuit is extremely simple. Each bit in the BCD code is set to a logic 1 by several decimal digits. Thus, to encode the decimal number into a BCD number, all of the decimal digits that produce a logic 1 for a particular bit position are ORed together. The circuit shown is a code converter for 8421 BCD; however, any BCD code can be formed in an identical manner.

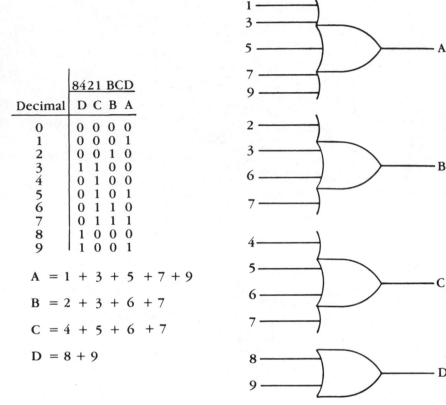

Decimal	8421 BCD D C B A
0	0 0 0 0
1	0 0 0 1
2	0 0 1 0
3	1 1 0 0
4	0 1 0 0
5	0 1 0 1
6	0 1 1 0
7	0 1 1 1
8	1 0 0 0
9	1 0 0 1

$$A = 1 + 3 + 5 + 7 + 9$$

$$B = 2 + 3 + 6 + 7$$

$$C = 4 + 5 + 6 + 7$$

$$D = 8 + 9$$

8-15   Decimal-to-BCD code converter.

Gray-to-Binary Converter

Two different methods for performing a conversion from gray code to binary are shown in FIG. 8-16. In a serial converter, a serial data stream is presented at the gray code input, most significant bit first. The binary output is an EX-OR of the current gray-code bit and the previous binary bit. This is another way of stating that if the current gray-code bit is a logic 0, the current binary bit will be the same as the previous binary bit. If the current gray-code bit is a logic 1, the current binary bit will be the complement of the previous binary bit.

With each clock pulse, a new gray-code bit is presented at the input and a new binary bit is generated at the output. As has been previously mentioned, one disadvantage to serial operations is the large number of clock periods required to perform a given logic function. This is also true for the serial gray code converter.

An alternate approach to gray code conversion is a parallel code converter. In this circuit, all bits are present at the input, and in a manner of

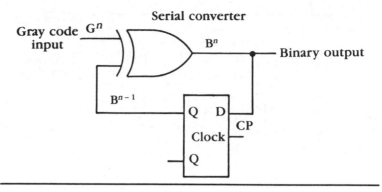

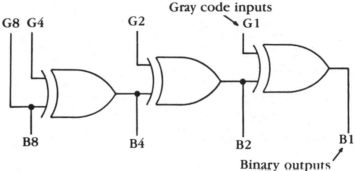

8-16 Two versions of the gray-to-binary converter.

nanoseconds, all outputs are also available. The disadvantage to a parallel converter is that many logic gates are required to perform a function that previously required only one gate.

BCD-to-Binary Converter

There are many ways of performing a BCD-to-binary conversion. Two commonly used methods are shown in FIG. 8-17. The serial converter method is the simplest of the two but, as with other serial operations, it is also the slowest. A BCD number to be converted is preset into a BCD *down counter,* while at the same time a binary *up counter* is reset to the all-0 state. At the next clock pulse, the clock control flip-flop is set and clock pulses are allowed to pass through the AND gate to the two counters. At each following clock pulse, the BCD counter will count down in a reverse binary sequence towards zero.

Meanwhile, the binary counter will count up (away from zero) at each clock pulse. The number of clock pulses for the BCD counter to reach count zero is equal to the BCD count that was preset into the counter. However, since the binary counter is also counting the same number of

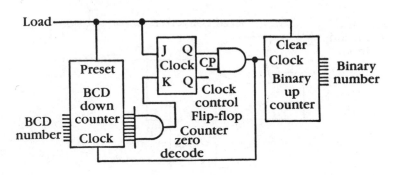

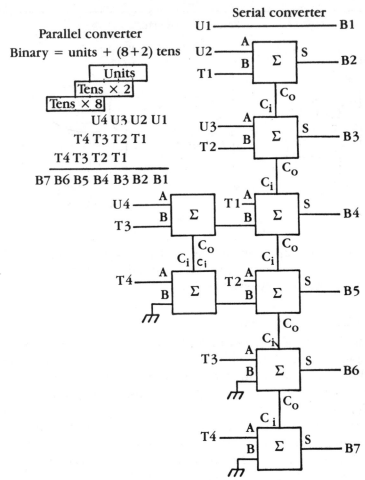

8-17 Two versions of the BCD-to-binary converter.

clock pulses, the correct binary count will continually be maintained by the binary counter. When the BCD counter reaches count zero, the count-zero decode gate resets the clock control flip-flop, thus discontinuing the count sequences of both counters. At this time, the numerical value contained in the binary counter can be read and it will be equivalent to the original BCD number.

A more complex method, which can perform the conversion in less than one clock period, is through the use of parallel adders. Any BCD number can be thought of as consisting of a *units* decade plus a *tens* decade, multiplied by a sum of powers of two that is equal to 10. If the units and tens are summed in an appropriate manner, the result will be a binary number consisting only of power of two. As an example, consider the conversion of the BCD representation for 69 to a comparable binary representation. The BCD representation is $T4T3T2T1\ U4U3U2U1 = 0110\ 1001$. Performing the summations described produces the results shown below:

			U4	U3	U2	U1
		T4	T3	T2	T1	
T4	T3	T2	T1			
B7	B6	B5	B4	B3	B2	B1

$\rightarrow$

			1	0	0	1
		0	1	1	0	
0	1	1	0			
1	0	0	0	1	0	1

The arithmetic is the same as that performed by the parallel converter circuit. A similar converter could be made up for a three-decade BCD number by recognizing that binary = units + tens $(8 + 2)$ + hundreds $(64 + 32 + 4)$. It can be seen, however, that the more decades in the BCD number, the larger the number of adders required. As a result, the technique shown is normally not useful beyond three BCD decades.

◆ Summary

There are several ways in which flop-flops may be combined to perform specific tasks. A *shift register* is a combination of flip-flops which may be used to store or transfer data between serial and parallel form. A *counter* is a device whose output increases or decreases in time with a clock input. This clock input may occur at all flip-flops simultaneously, creating a synchronous counter, or the output of one flip-flip may clock the next, in an asynchronous or ripple counter. External gates allow the same counter to be used to count to different maximum values. A *decoder* is a device, sometimes an AND gate, which looks for a specific set of inputs and creates a single output for that set of inputs. The opposite of a decoder is an *encoder,* which takes a single input and creates a corresponding set of outputs. A *code converter* is a group of gates that takes a set of bits and translates it to another set of bits.

Questions

1. What is a shift register? What are some of the different functions it can perform?

2. What is a counter? What is the difference between a sychronous and an asynchronous counter?

3. What is a decoder? How can an AND gate be used for this purpose?

4. What is an encoder?

5. What is a code converter? When might a code converter be used?

Problems

1. The circuit in FIG. 8-2 is to be used to convert 4 bits of data from parallel to serial form. What signals need to be present at the different inputs for this to occur? How many clock pulses must occur until the register is completely cleared?

2. Repeat Problem 1, except that 4 serial bits of data are to be converted to a 4-bit parallel word.

3. Using 3 JK flip-flops and any other necessary gates, draw a logic diagram for a modulo six synchronous counter.

4. Using 4 JK flip-flops and any other necessary gates, draw a logic diagram for a modulo twelve synchronous counter.

COMPLEX LOGIC NETWORKS

AFTER YOU COMPLETE this chapter, you will be able to describe different medium-scale and large-scale integrated circuits, such as:

☐ Up-down counters
☐ Multiplexers and demultiplexers
☐ Parity checkers
☐ Half and full adders

Prior to the advent of integrated circuits, digital logic was built from individual transistors and resistors. Thus, these were the components with which the designer and maintenance man were familiar. Later, many digital systems were built using integrated circuits, manufactured such that several gates or flip-flops were contained in one package. In these systems, the concern was with the individual logic functions.

As integrated circuit technology has grown, greater packaging densities have been achieved, and with these higher densities, it has become feasible to include complex functions in a single package. The packaging density is considered *medium-scale integration* (MSI) if between 25 and 100 gates are contained in a single package. *Large-scale integration* (LSI) encompasses packaging densities where hundreds of gates are included in a single package.

It is the purpose of this chapter to acquaint the reader with some of the complex functions currently included in modern digital systems. It should be remembered that there are many different logic families and many manufacturers of integrated circuits. Thus, no effort is made here to precisely represent any specific implementation of a complex function, but rather to show the various types of functions likely to be encountered.

COUNTERS AND SHIFT REGISTERS

Counters are among the most common complex logic functions available. Typical of the configurations encountered are binary and BCD counters operating either synchronously or as ripple counters. These counters are most frequently found in the DTL, TTL, ECL, and MOS logic families. An example of the advantages that can be obtained with MSI counters is shown by a standard up–down binary counter.

A truth table for a four-stage up–down counter is given in TABLE 9-1. From this truth table, J and K terms for individual flip-flops could be derived and a counter built out of gate and flip-flop packages. Such a circuit would require many packages to implement; and even then, if a failure occurred, detailed troubleshooting procedures would be required to isolate the faulty component. As a complex function, the entire circuit is contained in one package with a single line controlling whether the counter counts up or counts down. No external logic is necessary. Furthermore, if a failure should occur, it is merely necessary to determine that the entire counter is not functioning properly and to replace the counter package.

Table 9-1.   Truth Table for Four-Stage Up/Down Binary Counter.

Present State	Next Stage Count Up	Count Down
0000	0001	1111
0001	0010	0000
0010	0011	0001
0011	0100	0010
0100	0101	0011
0101	0110	0100
0110	0111	0101
0111	1000	0110
1000	1001	0111
1001	1010	1000
1010	1011	1001
1011	1100	1010
1100	1101	1011
1101	1110	1100
1110	1111	1101
1111	0000	1110

Also, because more functions can be included in the package, a very common feature is the inclusion of preset inputs for each individual flip-flop. With these inputs, it is a simple matter to program a given counter to count any desired modulo. For example, two different *modulo twelve* counters are shown in FIG. 9-1. Both are made up from modulo sixteen counter packages, with four counts skipped. If a binary count is desired, the recommended technique, shown in A, is to decode the end count of

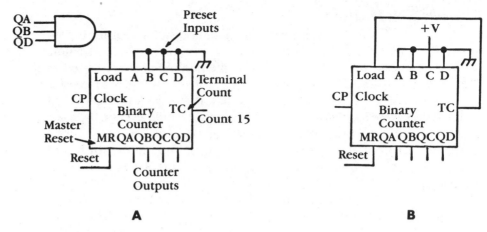

9-1 Two different modulo-twelve counters.

that particular modulo sequence, then preset the counter to the all-0 state. This simple change requires one gate.

A second method, shown in B, is by using the terminal count output that is often provided. In the example shown, the terminal count is defined as count 15. By using the terminal count to preset the counter, any number of the earlier states can be skipped. Here, states 0, 1, 2, and 3 are skipped. The disadvantage to this arrangement is that the count does not start at 0000 and progress in a normal binary fashion. Thus, any decodes taken from the counter must take this into account. Say, for example, that the second count of the sequence shown is to be decoded. State five (0101) would be the decode necessary to obtain the proper output.

Static Shift Registers

The register configurations described in chapter 8 can all be referred to as static registers. That is, they are all made up from flip-flops that can hold data indefinitely. Similarly, shift registers are static shift registers if clocks can be stopped without losing any data. Static shift registers can be restarted and the stored data can then be shifted out, or new data entered. All of the static register types are made in MSI configurations. The registers are usually 4, 5, 8, or 16 bits long and can be of the shift register type or the storage register type.

Almost any combination of serial–parallel entry or serial–parallel readout is available. Usually, the longer registers are of the serial-in–serial-out type. This is simply due to the fact that too many input and output leads would be required to make all of the intermediate stage inputs and outputs available.

Dynamic Shift Registers

If the clocks to a shift register must run continuously, the register is considered a dynamic shift register. Such is the case for the MOS dynamic storage described in chapter 5. Recalling, the method of storage used in the gate capacitance of a MOS inverter stage, using a MOS gate to transfer the stored charge from one stage to another. If the clock to this type of shift register were stopped, the capacitive storage would eventually decay and data would be lost. Very large amounts of delay can be obtained through the use of MOS dynamic shift registers. The number of bits of delay runs from about 25 to more than 1000 bits. This is obviously large-scale integration.

A two-stage dynamic shift register is shown in FIG. 9-2. A two-phase clock is used so that data can be transferred a half-stage at a time. This permits data to be copied into one half-stage without destroying the data contained in the other half-stage. The phase one clock can be considered an *intermediate transfer* clock, while the phase two clock can be considered the *read in–read out* clock.

Assume that an input data bit is present at the gate to the first inverter. When the phase one clock pulse occurs, the second inverter gate capacitance begins to charge. As soon as the phase one clock pulse goes away, the charge is stored in the second inverter of the first stage. Thus, the phase one clock has temporarily transferred the data within the first stage. When the phase two clock pulse occurs, the second-stage inverter gate capacitance charges.

Now, when the phase two clock is removed, the data is stored in the second stage. Hence, the phase two clock has read the data out of the first stage and into the second stage.

MULTIPLEXERS

A digital multiplexer selects one signal from among several inputs and applies this signal to the output. The selection of the appropriate signal is controlled by an *input select code*. One typical use for multiplexers is the sequential (serial) application of a number of signals onto a single line. If this process is strictly a function of time (that is, if each line is sequentially selected), the process is referred to as *time division multiplexing*. Functionally, time division multiplexing might be considered another method of parallel-to-serial conversion.

Another use for digital multiplexers is data routing. Suppose, for example, that one signal is desired in a particular operating mode, while some other signal is desired if the mode is changed. The multiplexer provides the means for making this data selection.

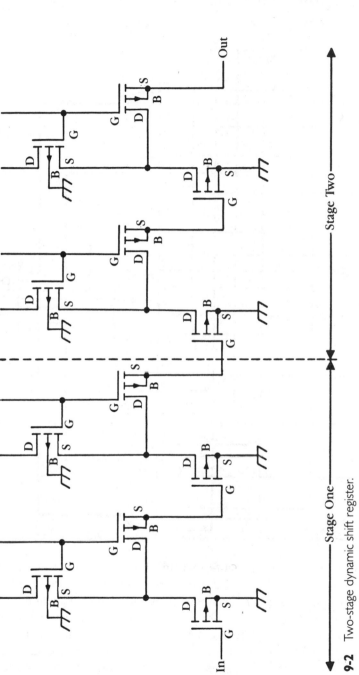

9-2 Two-stage dynamic shift register.

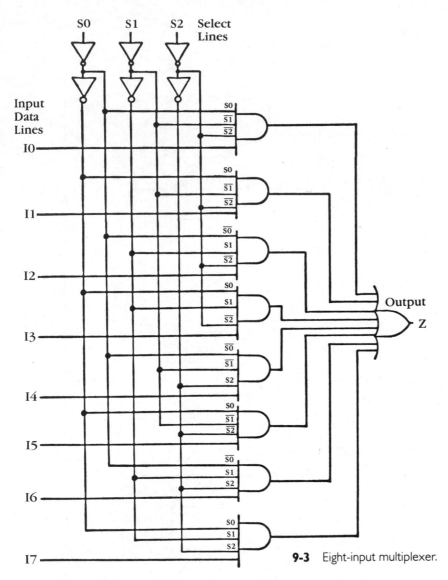

9-3   Eight-input multiplexer.

An example of an eight-input multiplexer is shown in FIG. 9-3. The entire logic array is simply a group of AND gates that decode the select lines and apply the appropriate input signal to an OR gate at the output. A number of inverters are included to obtain the complements of the select lines and to provide buffering. Of course, the reason for inclusion in this chapter is that the multiplexer is available as a complex function in the MSI class of integrated circuits. As a complex function, it is observed that many interconnections and individual gates are combined into a single package, thus simplifying the design and maintenance functions.

DEMULTIPLEXERS

A demultiplexer basically reverses the multiplexing function. It takes data from one line and distributes it to a given number of output lines. Figure 9-4 shows a one-line-to-four-line demultiplexer circuit. The input data line

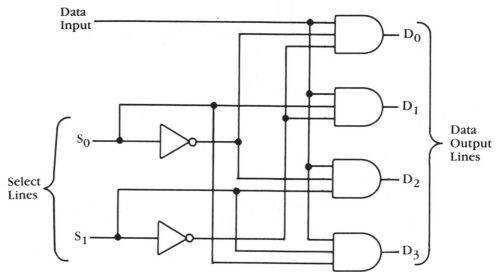

9-4 A one-line-to-four-line demultiplexer.

goes to all of the AND gates. The two select lines enable only one gate at a time and the data appearing on the input line will pass through the selected gate to the associated output line.

PARITY GENERATORS/CHECKERS

To check for or generate the proper parity in a given code word, a very basic principle can be used: The sum (disregarding carries) of an even number of 1's is always 0, and the sum of an odd number of 1's is always 1. Therefore, to determine if a given code word is even or odd parity, all of the bits in that code word are summed. OR gates can be used to sum these inputs (see FIG. 9-5). (Recall the discussion of error-detection in chapter 2.)

The parity generation/detection logic for a four-bit code (including parity) is shown in FIG. 9-6A and that for an eight-bit code is a FIG. 9–6B.

ADDERS

Another common logic network is the *adder circuit.* There are usually two forms of the adder in common applications. The first is a *half adder,* which adds two binary bits and generates a sum and a carry output. The second

(A) Summing two bits.

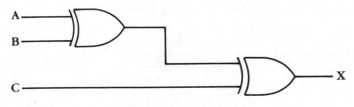

(B) Summing three bits.

9-5 Parity generator/checker using OR gates to sum the inputs.

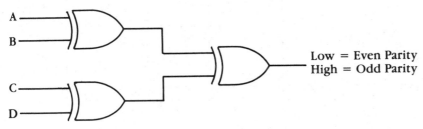

Low = Even Parity
High = Odd Parity

(A) Four-bit parity checker

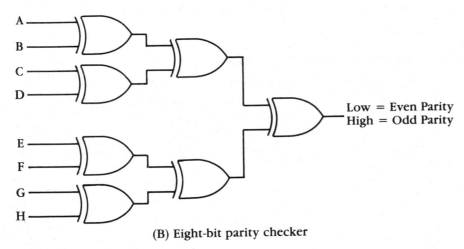

Low = Even Parity
High = Odd Parity

(B) Eight-bit parity checker

9-6 Parity generators/checkers.

form is the *full adder,* which adds two binary bits plus a carry input to produce the sum and carry outputs. The two circuits are shown in FIG. 9-7.

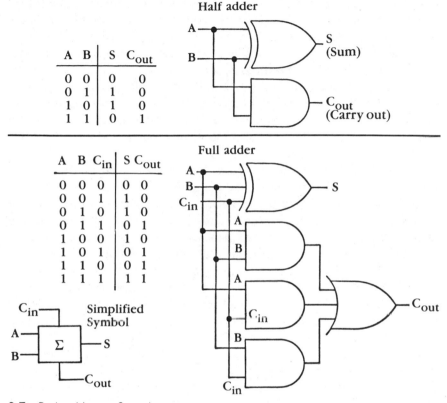

Half adder

A	B	S	C_{out}
0	0	0	0
0	1	1	0
1	0	1	0
1	1	0	1

Full adder

A	B	C_{in}	S	C_{out}
0	0	0	0	0
0	0	1	1	0
0	1	0	1	0
0	1	1	0	1
1	0	0	1	0
1	0	1	0	1
1	1	0	0	1
1	1	1	1	1

9-7 Basic adder configurations.

Consider first the half adder circuit. The truth table shows that this is the standard binary addition first identified in chapter 3. The sum is simply an EX-OR function of inputs A and B and a carry is generated if two 1's are added.

But the situation is not all that simple if more than two bits are to be added. In this case, there is a possibility of a carry input to the addition from a previous bit. Hence, binary addition becomes a three-variable problem, with outputs as shown in the full adder truth table. It is seen that the sum output is the EX-OR of all three inputs, while a carry bit is generated any time more than one of the inputs is a logic 1.

Instead of drawing all of the logic each time an adder is used, the simplified symbol shown will be employed. If the symbol has a carry input, then a full adder is implied; otherwise, a half adder can be used.

Serial Adder

A block diagram for a serial adder is shown in FIG. 9-8. Some binary number, A4A3A2A1, is stored in register A. A second number, B4B3B2B1, is stored in register B. The process of addition begins by resetting the D flip-flop so that the carry bit will be a logic 0 to start with. The two registers now shift out their contents, both at the same time, while register C is simultaneously shifting in sum bits as they are generated. For example, suppose that the two numbers to be added are as shown in the diagram. At the first clock pulse, the inputs A and B are both logic 1 and there is no carry in. Therefore, the first sum bit, a logic 0, is shifted into register C and a logic 1 carry bit is stored in the D flip-flop. At the second clock pulse, the carry bit will be present at the carry input and inputs A and B will be 0 and 1, respectively.

According to the truth table rules, the sum bit will again be a logic 0 and another logic 1 carry bit will be generated. Following this procedure through the fourth clock pulse results in a logic 1 carry bit being left in the D flip-flop and no further data bits left in the A and B registers. But the carry bit is a perfectly valid part of the sum; therefore, one last clock pulse must be used to complete the serial addition. This example demonstrates that, when adding two binary numbers, the sum register must always be able to accommodate one more bit than was in the input registers.

Registers A, B, and C are normally existing registers, used for multiple functions in the system, and the only extra logic required, in addition of this type, is the full adder circuit and a single flip-flop.

The main advantage to a serial adder is its simplicity and the minimum amount of logic required. Its primary disadvantage is that it takes many clock periods to complete one addition; so if there are many such additions to be performed, the time required may become excessive. A parallel adder can perform this same operation in less than one clock time.

Parallel Adder

The parallel adder shown in FIG. 9-12 uses the same full adder circuit as the serial adder, but adds all bits at the same time. As can be seen, this configuration requires one complete full adder circuit for each bit position. If large numbers are being added, this represents a considerable amount of logic. Further, the parallel adder still requires registers A, B, and C in which to store its results.

The big advantage to parallel adders is that the entire operation can be performed in nanoseconds (thousandths of microsecond) as compared to microseconds (millionths of a second) for the serial adder.

The same example as was used for the serial adder can be used to

Serial Addition

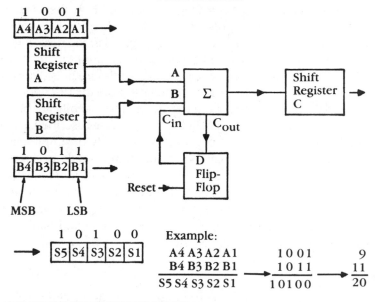

Example:

$$
\begin{array}{l}
A4\ A3\ A2\ A1 \\
B4\ B3\ B2\ B1 \\
\hline
S5\ S4\ S3\ S2\ S1
\end{array}
\longrightarrow
\begin{array}{l}
1\ 0\ 0\ 1 \\
1\ 0\ 1\ 1 \\
\hline
1\ 0\ 1\ 0\ 0
\end{array}
\longrightarrow
\begin{array}{l}
9 \\
11 \\
\hline
20
\end{array}
$$

Parallel Addition

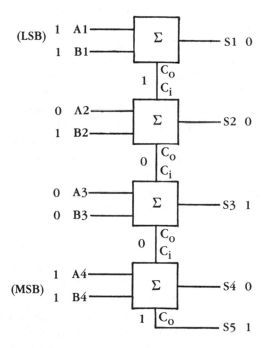

9-8   Serial and parallel adders.

analyze the parallel adder circuit. The two inputs A1 and B1, as before, are each at logic 1, resulting in S1 = 0 and a carry of 1. At the same time, inputs of A2B2 = 01 are also present at the second adder. This generates a sum S2 = 0 and again a logic 1 carry is propagated to the next stage. The action described is repeated for each parallel stage until the final carry is obtained. The maximum speed of operation for this circuit is determined by the length of time it takes a carry to propagate from the first adder (least significant bit) to the carry out of the last adder (most significant bit).

Adder circuits such as the parallel full adder just described are naturals for medium-scale integration. Consisting entirely of simple gate circuits, a typical 4-bit full adder can be included in a single package. The main limiting factor with MSI adders is the number of input and output leads required. In the case of a 4-bit full adder, there are 8 inputs for the bits to be added, 4 outputs representing the sum bits, a carry input, and a carry output. For this simple circuit, there are 14 input and output signal leads required, not including power and ground.

SUBTRACTORS

Subtractors, like full adders, are easy to implement as medium-scale integrated circuits. Actually, a subtractor is usually built from a full adder by adding inverters on the inputs to the number being subtracted. As was mentioned in chapter 1, two's complement notation is extremely useful for addition and subtraction of binary numbers. Use of the full adder as a subtractor is a good example of this concept.

Recalling, two's complement subtraction was performed by inverting all of the bits in the number being subtracted, adding one, then adding the two numbers together. This is exactly what is done in FIG. 9-9. Although the inverters are shown external to the full adder, a subtractor package would include the inverters inside the unit. Several examples are shown to demonstrate the two's complement technique of subtraction. These examples show positive numbers being subtracted from one another; however, the circuit works equally well for negative numbers, provided that they are in two's complement notation.

COMPARATORS

A digital comparator circuit is merely an extension of the use of adder and subtractor circuits. Assuming there are two numbers, A and B, to be compared, the technique consists of subtracting B from A, then noting whether the result is positive, negative, or zero. If $A - B$ is positive, then A is greater than B. Similarly, if $A - B$ is negative, then B is greater than A. The condition where $A - B = 0$ is decoded to show that $A = B$.

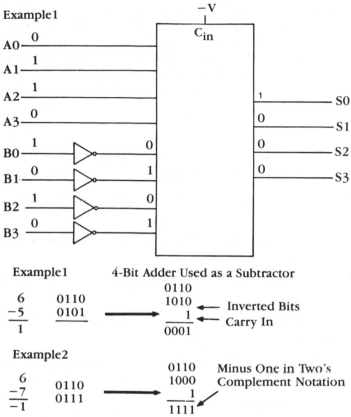

9-9 Subtractor using two's complement arithmetic.

The comparator circuit has the same type of limitations as adders and subtractors. The number of bits that can be compared in a single package is strictly a function of the number of leads that can be made available in a given package configuration.

Summary

Groups of gates may be combined into single packages called *medium-* (25–100) or *large-* (>100) scale integration. Some examples of these are the *up-down counter, shift register, multiplexer, demultiplexer, parity generator/checker, adder, subtractor,* and *comparator.* The multiplexer takes signals from several sources and, depending upon one or more inputs to select lines, determines which of the inputs is to appear on the output line. Conversely, the demultiplexer takes a single bit of data and, depending upon the inputs to select lines, places that data onto one of several output lines. A parity checker is used to determine whether odd or even parity exists, and it can also be used to generate the required parity bit. *Half adders* are used to add two bits, while full adders also include an input carry. A *full adder* can also be used to perform the subtraction operation. A comparator is used to determine which of two numbers is larger.

Questions

1. What is an up-down counter? What advantage does it have over an up-only counter?

2. What is the difference between a static shift register and a dynamic shift register?

3. What is a multiplexer? How does it work? Give one example when it might be used.

4. What is a demultiplexer? How does it work? Give one example when it might be used.

5. What is parity checking? What is the primary gate used in parity checking?

6. What is a half adder? A full adder?

7. Explain the difference between the operation of a parallel adder and a serial adder.

8. What needs to be done to use a full adder as a subtractor?

9. What does a comparator do?

Problems

1. In FIG. 9-1A, what changes would be necessary to convert this to a modulo ten counter?

2. Repeat Problem 1 to obtain a modulo six counter.

3. In FIG. 9-3, data from which input line appears on the output for $S0 = 0$, $S1 = 1$, and $S2 = 1$?

4. In FIG. 9-3, data from which input line appears on the output for $S0 = 1$, $S1 = 1$, and $S2 = 0$?

5. In FIG. 9-4, data from the input appears on which output when S0 = 0, and S1 = 1?

6. In FIG. 9-4, data from the input appears on which output when S0 = 1, and S1 = 0?

7. Show how the parity checker in FIG. 9-6A can be used to add an odd parity bit to word DCBA.

8. Show how the parity checker in FIG. 9-6A can be used to add an even parity bit to word DCBA.

9. Show how two 4-bit parallel adders can be combined to create an eight-bit parallel adder.

10. Use logic gates to compare the 2-bit numbers A1A0 and B1B0. One of the three outputs should be high: $A = B, A > B, A < B$. Hint: Write a logic expression for each of the three possible outputs first.

MEMORIES

AFTER YOU COMPLETE this chapter, you will be able to:

☐ Describe the basic categories of memories
☐ Understand the difference between serial data access and parallel data access
☐ Explain how a PLA can be programmed to create specific outputs

Registers, discussed in chapter 8, are a type of storage device. Registers are normally considered temporary storage devices, whereas the term *memory* is typically used for a device that holds data for longer-term storage. Also, a typical memory can hold a much larger amount of data than a typical register.

Modern applications of data processing systems require that huge amounts of information be permanently stored and readily accessible. Banking, inventory control, the census, and social security are just a few examples where a great deal of information must be kept and processed.

There are two basic categories of memories in current use: *semiconductor* and *magnetic*. Within each category are a variety of memory types. Generally, the semiconductor memories are used for smaller capacity and faster access requirements. The various types of magnetic memories are used for larger capacity storage, but it generally takes longer to access the information.

The semiconductor memories discussed in this chapter are *random-access memory* (RAM), *read-only memory* (ROM), *programmable read-only memory* (PROM), *read-mostly memory* (RMM), *content-addressable memory* (CAM), *programmable logic array* (PLA), *first in-first out memory* (FIFO), and the *last in-first out memory* (LIFO). The magnetic medium discussed is the *magnetic bubble memory* (MBM).

Semiconductor memory arrays are of two basic types, *serial data access* and *parallel data access*. Static and dynamic shift registers are examples of serial memory organizations. Storage registers and decoding matrixes are examples of parallel memory. Often, memory requirements become quite large, such that large-scale integration techniques are justi-

fied. Using modern technology, relatively sophisticated data entry and re-
trieval schemes can be included in a single integrated-circuit package. The
descriptions that follow describe memory arrays with up to several-
thousand-bit capacities, which are typically found in a single package.
However, larger memories can always be formed by interconnecting a
number of such packages.

RANDOM-ACCESS MEMORIES

A random-access memory (RAM) is a read/write memory consisting of a
number of memory cells which can be randomly accessed, based upon an
input address. For large-scale integration, the memory cells are usually
cross-coupled multi-emitter transistor flip-flops with appropriate input
and output select logic added to perform the random-access function. This
random-access memory is a volatile type, because all data is lost if power
is removed.

A schematic diagram of a basic RAM cell is shown in FIG. 10-1. To
understand the operation, assume initially that the flip-flop is in some
given state and that the address select line is low. The two transistors in
this case look exactly like a standard common-emitter flip-flop with the
emitters tied to ground. Clearly, this flip-flop will store a logic 1 or a logic
0 indefinitely, so long as power is not removed.

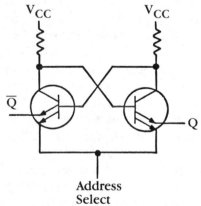

10-1   Bipolar RAM cell.

Next, assume that at some later time the address select line is set high.
Then, the two output emitters Q and $\overline{Q}$ will independently have current
flow or not, depending on which transistor is conducting and which is cut
off. Thus, to read a bit from the flip-flop, it is only necessary to sense which
emitter has current flow to determine the state of the flip-flop.

Finally, if it is desired to write new data into the flip-flop, the address select line is again set high so that the two emitters are independent. Now, if a logic 0 is placed on one emitter and a logic 1 on the other, the transistor with the lower emitter voltage will conduct and the opposite transistor will cut off. Hence, a logic 0 will be stored by the transistor which has a logic 0 applied to its emitter, while a logic 1 will be stored by the transistor which has a logic 1 applied to its emitter.

To understand how the individual memory cells make up a complete memory array, a block diagram for a typical random-access memory is shown in FIG. 10-2. This memory contains 64 bits of storage, arranged in

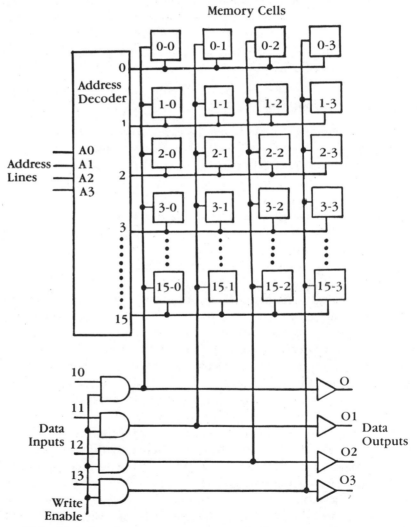

10-2   Typical random-access memory organization.

memory cells numbered by row and column. There are 16 rows, each containing 4 memory cells. Each row can be written into or read out from individually. To write data into the memory, a *write enable* signal is required, in addition to 4 data bits at the data input lines. The address specified at the address input lines defines the memory cells that will receive the input data. The read function is performed by setting the write enable line low and applying an address indicating the row of memory cells to be read out.

Typical usage of a random-access memory requires two separate address generators, one for writing and one for reading. Data is written into known locations in the memory as designated by the written address generator. Hence, when data is retrieved from the memory, it is accessed by knowing in advance the memory location where the data resides and addressing this location. As will be shown, other memory types do not require that the specific memory address be known.

READ-ONLY MEMORIES

A *read-only memory* (ROM) is simply a very large array of gates wired such that upon application of an address consisting of several bits, a fixed set of outputs is obtained. The array is a memory because it stores a known set of outputs for a given input condition. The memory is known as *read-only* because no new data can be written into it.

The memory is wired in a fixed manner. Read-only memories are very useful as lookup tables to perform code conversions or to look up mathematical functions. Typical mathematical functions include sine, cosine, or logarithmetic lookups or even binary multiplication tables.

To demonstrate the basic simplicity of the read-only memory concept, a simplified schematic of a diode memory, wired to perform decimal-to-binary conversion, is shown in FIG. 10-3. This circuit has memory, because stored in the logic array (the diodes plus wiring) is a binary number for every decimal number that can be applied at the input. The inputs are decoded in a standard manner, forming four bits representing the powers of two that correspond to the decimal number applied.

For each output line, a diode connection is made where a logic 1 output is desired and no connection is made where a logic 0 output is desired. As can be seen, this read-only memory represents a logic circuit consisting of AND and OR gates whose truth table provides outputs corresponding to the binary bits that directly relate to the decimal inputs.

As a matter of fact, every read-only memory is defined in terms of its truth table. The table shows what outputs are obtained for every given input configuration. Thus, to know the truth table is to know the function of any given read-only memory.

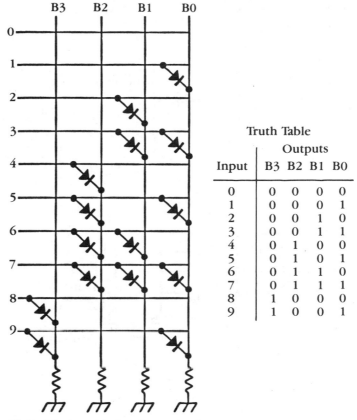

10-3   Diode matrix forming read-only memory.

The diode memory of FIG. 10-3 can be manufactured in several different ways. The illustration shows a memory that might be made of discrete components, using diodes only in those locations where needed. As an alternative, the read-only memory might be manufactured as an integrated circuit containing diodes at all possible junctions. The actual programming of the read-only memory, then, would be performed by the unique metal interconnection mask for that particular integrated circuit. Then, the 1's and 0's for each particular bit location are determined by connecting or not connecting the diode in that location.

The diode memory shown in FIG. 10-3 is fine for use where only one input line at a time will be enabled. However, most ROMs are designed to produce specified outputs for every combination of inputs, where there is no restriction on the states the input lines may assume.

A very simple example of such a ROM is shown in FIG. 10-4. This ROM has two address variables, X and Y. For each combination of X and Y, there

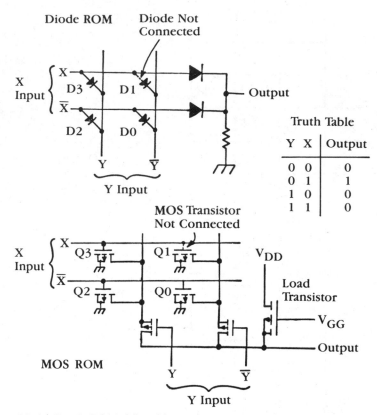

Truth Table

Y	X	Output
0	0	0
0	1	1
1	0	0
1	1	0

10-4 Simple ROM, fully addressed.

is a defined output; either a 1 or a 0. Where a 1 output is desired, the diode or transistor at the appropriate junction is not connected. Where the connection is made, a 0 output is obtained for the state that corresponds to that intersection. The ROM shown is trivial because it is clear that a two-input gate could perform the same function. But, conceptually, the same idea is applied to ROMs with many input variables and more than one output.

There are two versions of the same functional ROM shown to illustrate that MOS transistors as well as bipolar elements can perform the desired memory function. Note that both versions are of the type normally manufactured as integrated circuits. That is the 1's and 0's are programmed into the ROM by connecting or not connecting the individual logic diodes or transistors.

As a final example of the type of memory obtained with the ROM arrangement, FIG. 10-5 shows a complex array in block diagram form. This array has 16 different address locations for each of 4 output lines. Hence, it is sometimes referred to as a 64-bit ROM. This particular memory is

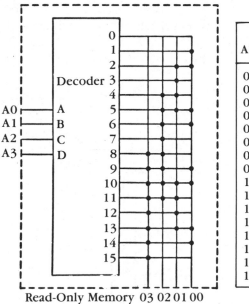

ROM Truth Table

Inputs				Outputs			
A3	A2	A1	A0	Q3	Q2	Q1	Q0
0	0	0	0	0	0	0	0
0	0	0	1	0	0	0	1
0	0	1	0	0	0	1	1
0	0	1	1	0	0	1	0
0	1	0	0	0	1	1	0
0	1	0	1	0	1	1	1
0	1	1	0	0	1	0	1
0	1	1	1	0	1	0	0
1	0	0	0	1	1	0	0
1	0	0	1	1	1	0	1
1	0	1	0	1	1	1	1
1	0	1	1	1	1	1	0
1	1	0	0	1	0	1	0
1	1	0	1	1	0	1	1
1	1	1	0	1	0	0	1
1	1	1	1	1	0	0	0

Read-Only Memory 03 02 01 00

10-5   Binary-to-gray code conversion using ROM.

connected to perform a binary-to-gray code conversion. The actual bit connections are represented in the figure by a dot at the appropriate inter section for a logic 1 and no dot for a logic 0. The versatility of ROMs is that, given a number of input variables, the output lines can be programmed to give any conceivable pattern of 1's and 0's.

PROGRAMMABLE READ-ONLY MEMORIES

ROMs, as purchased from an IC manufacturer, are normally already pre-programmed to perform some standard table lookup function. If a large quantity of a particular memory configuration is desired, the manufacturer will fabricate the unit to the customer's requirements. However, a given digital system may require some unique table lookup that is not an off-the-shelf product and only one or two of the memories are needed. In this case, it is not economically feasible for the manufacturer to tool up for the necessary production cycle. As an alternative, programmable read-only memories can be purchased that have all of their programming inter-connections made with a fusible material.

To program such a read-only memory, a one-time write process is performed that essentially open-circuits all of the diodes where no connection is desired. In this way, the user can make any memory function needed. Programmable read-only memories are frequently found in one-of-a-kind digital systems.

An example of a PROM is shown in FIG. 10-6. This memory is a bipolar memory as opposed to the MOS memory because it uses bipolar transistors. A PROM can just as easily be made in the MOS version, but its operating speed is typically somewhat slower than the bipolar version.

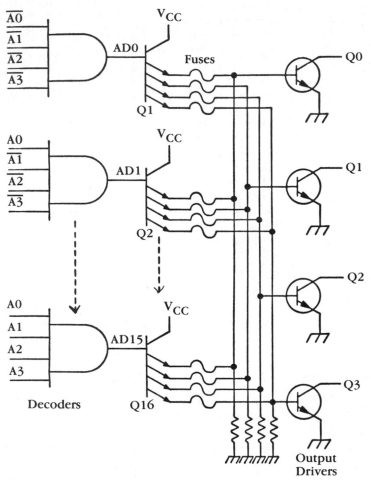

10-6 Bipolar programmable ROM.

Referring to the figure, the 16 addresses (AD0 through AD15) are decoded in a standard logic fashion to apply an enable voltage to one of 16 multi-emitter transistors with "fuses" connecting the emitters to the output driver stages. There are four outputs labeled 00 through 04, each output requiring a separate driver transistor. To program this memory, the desired address is selected, then a fairly large negative collector voltage is applied to the output being programmed. As a result, the transistor base-collector junction becomes forward biased and draws enough current to cause the selected "fuse" to open, thereby programming a logic 1 in that location. If the fuse is left unblown, a logic 0 is retained at the location.

READ-MOSTLY MEMORIES

There are certain instances when a particular set of lookup tables may be needed for weeks or months. This is definitely an application for a read-only memory. Occasionally, however, the user would like to modify the data in these tables to meet his changing needs. Here, a read/write memory would be preferred; but a standard read/write memory cannot be used: it uses flip-flops for data storage, and flip-flops lose their stored information when power is removed and reapplied.

One solution to the problem is a read-mostly memory (RMM). In the logic circuit this type of memory is used exactly like a read-only memory. The difference is that the read-mostly memory can be repeatedly reprogrammed to give new sets of outputs. Instead of the fusible material in programmable read-only memories, the read-mostly memory uses transistors with metal nitride-oxide (MNOS) as the semiconductor element for storage of data bit patterns. Storage is accomplished by electrically altering the threshold voltages of the individual transistors that make up the memory.

Physically, the MNOS transistor like the one shown in FIG. 10-7 is quite

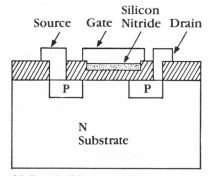

10-7 NMOS transistor.

similar to an ordinary MOS transistor. The mechanism for varying its threshold voltage lies in the ability to trap electrical charges in the silicon nitride gate insulator. If a positive charge is trapped, the threshold voltage of the transistor is lowered. Similarly, if a negative charge is trapped, the threshold is raised. A low-threshold transistor (one that will turn on) represents a logic 1, while a high threshold transistor (one that remains off) represents a logic 0.

When a number of MNOS transistors are combined into a memory array, the result of the trapped charges is to store logic 1's and 0's in a unique pattern to form a custom-programmed read-only memory. The terminology "read-mostly memory" is derived from the fact that, at least occasionally, the memory must be written into. But, for the most part, only the read function is performed.

The read-mostly memory is nonvolatile in that power can be removed from the transistors then reapplied as often as desired while the trapped charges representing the binary data will not be lost. After a very long period, perhaps as much as a year, the MNOS transistors tend to lose their charges; and therein lies their principal disadvantage. An MNOS memory array must be occasionally refreshed (reprogrammed) so that it will not lose the trapped charges.

CONTENT-ADDRESSABLE MEMORIES

The random-access memories just described depend upon prior knowledge of memory location to read out data. A content-addressable memory (CAM), however, reads out data by applying a known data set at the inputs then searching memory contents to see if there are any locations containing the same information. If there are, a *match* condition occurs and the addresses of all matching locations are made available. Normally, the content addressable memory will be organized such that some key portion of the data will be searched, while the rest of the data is ignored.

For example, a particular memory might contain frequency, pulse width, and amplitude information concerning a given set of pulses. A standard search function might be to scan memory to see how many pulses of a given frequency there are in memory. Hence, only the frequency portion of the data would be searched for a match condition. Once a match condition is obtained, the specific memory locations are known, and the entire data files can then be read out by direct memory addressing.

Writing into the memory is also done by direct address. The determination of a write address is based either on known empty memory locations or by searching for locations in which data is no longer needed, then writing in those addresses.

A simplified diagram of a 16-bit content addressable memory is shown in FIG. 10-8. There are four rows in the memory, each containing four memory cells. Each row has an individual address bit, allowing the memory cells to be searched, read, or written into.

Assume that the memory contains the data bits shown and a search operation is to be performed to determine if the number 5 (0101) is contained anywhere in the memory. If the address bits are all set to 0 and the search data (0101) is applied at the data input lines, a match condition will occur at all rows that contain the desired information. In this case, only

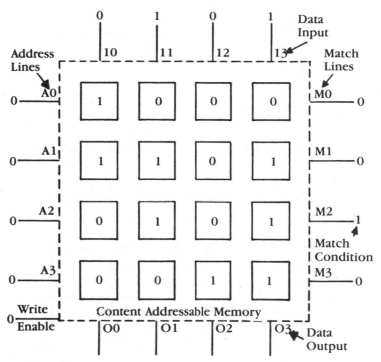

10-8 Search operation in content-addressable memory.

row 3 contains 5. If it is desired, row 3 can be read out by simply setting address bit $\overline{A2}$ to a logical 1. To write in the memory, the write enable line is set to a logic 1 along with the selected address line. The data present at the data input lines will then be written into the selected row of the memory.

PROGRAMMABLE LOGIC ARRAY

A programmable logic array (PLA) essentially consists of an array of AND-OR logic with inverters that can be programmed to produce desired logic functions on the outputs. An example of a small PLA is shown in FIG. 10-9. Each connection in the figure is *mask programmable,* which means that it is permanently programmed at the manufacturer from customer-specified instructions. Once the memory is programmed, it cannot be changed. This three-variable example is programmed to produce the indicated logic expressions by masking open certain connections.

A *field programmable logic array* is one that can be user programmed rather than mask programmed by the manufacturer. FIG. 10-10 shows a block diagram for this type of logic array.

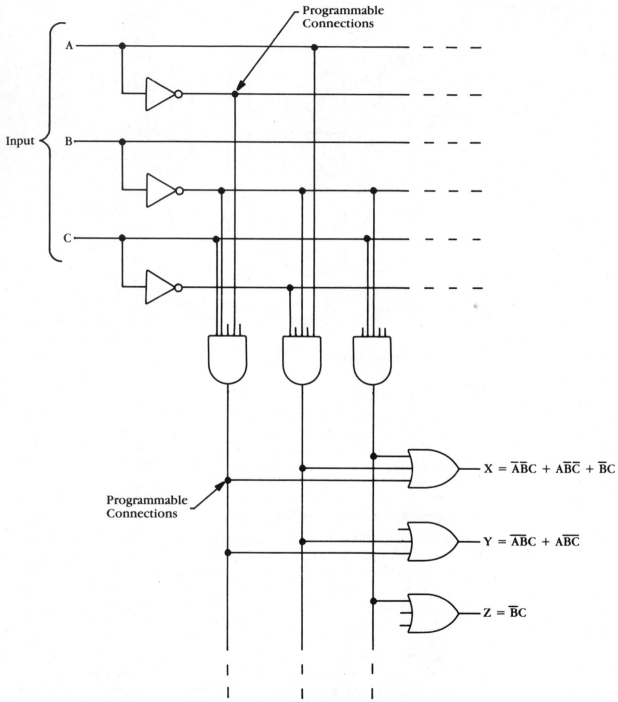

10-9 Simple example of programmable logic array.

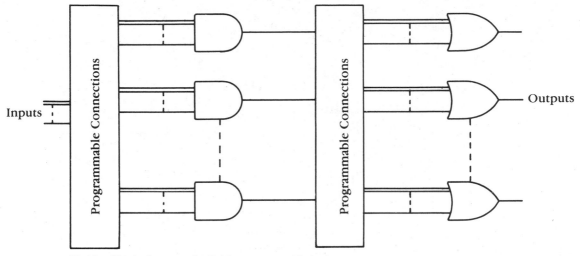

10-10 Block diagram of a field-programmable logic array.

FIRST IN–FIRST OUT MEMORIES

A modification of the random-access memory concept is the idea of a first in–first out (FIFO) memory. Also referred to as a *silo* memory, the analogy is quite appropriate. In a farmer's silo, the feed is put into the top (usually at quite a high speed) and is taken out more slowly at the bottom. The first feed into the silo goes directly to the bottom of the silo, thus also becoming the first feed out. The first in–first out memory receives data at one data rate and outputs data at some other unrelated data rate. The first data in, like the silo, becomes the first data out. A typical use for a first in–first out memory might be the interface between a high-speed digital computer and a much slower typewriter. Blocks of data can be fed into the first in–first out memory at high speed, until the memory is full. Then, the typewriter can read the same data out at a slower rate. When the memory is empty, the computer can again refill the memory cells at a high data rate.

A block diagram of a first in–first out memory is shown in FIG. 10-11. Recalling the random-access memory operation, separate read address and write address generators were commonly used to access specific data locations. With the first in–first out memory, the read and write addresses each increment sequentially so that two counters advance one count each time data is entered into the memory while the read counter advance a single count each time data is read out. Notice that, external to the memory, there is no need to know memory locations. Indeed, there is no provision to access data by specific location, even if this were desired.

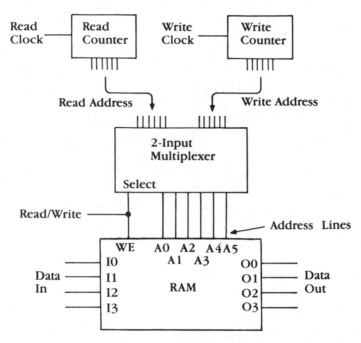

10-11   First in-first out memory block diagram.

LAST IN–FIRST OUT MEMORIES

The last in–first out (LIFO) memory is used in many applications involving microprocessors and other computing systems. It allows data bytes to be stored and then recalled in the reverse order; that is, the last data byte to be stored is the first byte to be retrieved.

A LIFO memory is commonly referred to as a *push-down stack*. In many microprocessor systems, it is implemented with a group of registers as shown in FIG. 10-12, known as a *stack*. A stack can consist of any number of registers, but the register that is the first to receive incoming data is the top of the stack.

A byte of data (eight bits) is loaded in parallel onto the top of the stack. Each successive byte pushes the previous byte "down" the stack into the next register. See FIG. 10-13A. Notice that the new data byte is always loaded into the top register and the previously stored bytes are pushed farther down the stack.

Figure 10-13B illustrates what occurs when data bytes are retrieved. Since the *last* byte entered is always at the top of the stack, it will be the *first* to be "pulled" or "popped" from the stack. Simultaneously, the other bytes move up into the next stack position.

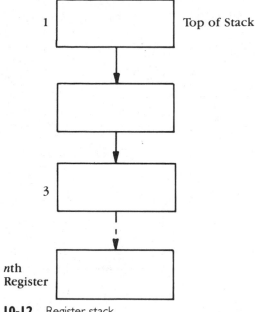

10-12   Register stack.

MAGNETIC BUBBLE MEMORIES

Magnetic bubble memories can be considered in some ways to be analogous to magnetic disk memories. In both types, data is stored as states of magnetization in a thin magnetic film. In an MBM, data bits are stored in the form of magnetic "bubbles" moving in thin films of magnetic material. The bubbles are actually cylindrical magnetic domains whose polarization is opposite to that of the thin magnetic film in which they are embedded.

When a thin wafer of magnetic garnet is viewed by polarized light through a microscope, a pattern of "wavy" strips of magnetic domains are visible. In one set of strips, the tiny internal magnets point up and in the other areas they point down. As a result, one set of strips appears bright and the other looks dark when exposed to the polarized light. This is illustrated graphically in FIG. 10-14A.

If an external magnetic field is applied perpendicular to the wafer and slowly increased in strength, the wavy domain strips whose magnetization is opposite to that of the field begin to narrow. This is illustrated in FIG. 10-14B.

At a certain magnitude of external field strength, all these domains suddenly contract into small, circular areas called *bubbles*. See FIG. 10-14C. These bubbles typically are only a few micrometers in diameter and act as

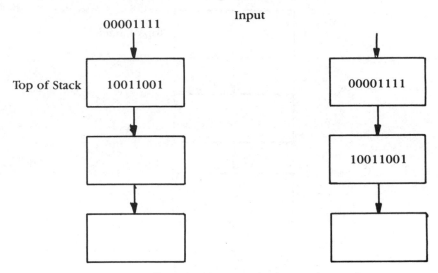

A Bytes being pushed onto the stack.

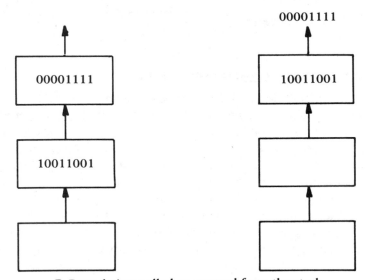

B Bytes being pulled or popped from the stack.

10-13   Illustration of how data is pushed and pulled (or "popped") to and from the stack.

tiny magnets floating in the external field. The bubbles can be easily moved and controlled within the wafer by rotating magnetic fields in the plane of the wafer, or by current-carrying conductive elements.

The presence or absence of a bubble represents a binary 1 or 0 respectively. In the simplest MBM arrangement, bubbles can be generated,

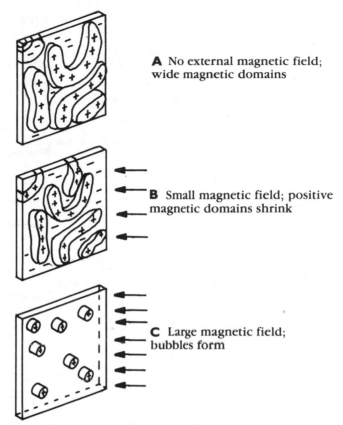

A No external magnetic field; wide magnetic domains

B Small magnetic field; positive magnetic domains shrink

C Large magnetic field; bubbles form

10-14   Creation of magnetic bubbles in a thin magnetic wafer by applying an external magnetic field.

shifted, and detected in an *endless-loop shift register.* The shift register is formed by a pattern of shapes that are deposited on the magnetic garnet wafer. These endless-loop shift registers provide the basis for mass data storage.

Basically, the data bits are stored in several minor loops and are transferred into a single major loop to be read or altered. Figure 10–15 shows a simplified diagram of the major and minor loop organization. To write data, bubbles enter into the major loop via a bubble generator from the reservoir loop (lower left corner in the figure) under control of a write command. Then, from the major loop, data bits can be transferred into a minor loop for storage. The rotating magnetic field keeps the bubbles moving at all times. Selective field application with the various control inputs is used to produce transfers from loop to loop.

To read data, bits transfer from a minor loop to a major loop. They

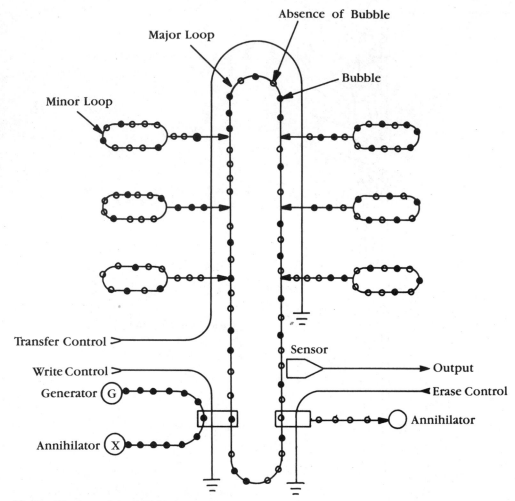

10-15   Diagram of the MBM loop concept.

are then sensed by the detector element (as to whether they are a 1 or a 0) and the result is the output data bits. Data can also be erased by transferring them from the major loop to the annihilator loop (lower right corner in the figure). We now have MBMs with a storage capacity of 1 million bits and higher, and that number is certain to grow.

Summary

Both registers and memory are storage devices. However, register is generally considered short-term, and most registers lose their data when power is turned off. Memory is considered long-term, and it retains its information when power is turned off. Memory is normally broken down into two categories: semiconductor and magnetic. Most memory is of the semiconductor type. Memory is classified according to how data is entered and retrieved and whether data can be written and read from it. *Random-access memory (RAM)* is read/write memory, and *ROM* is read-only memory. *PROM* is once-programmable, and *RMM* can be read and written to multiple times. *Content-addressable memory (CAM)* is unique in that data is accessed based on the contents of the memory, rather than on the address. A *programmable logic array (PLA)* is a form of a decoder in which the outputs have a fixed logic output. Memory registers can also be stacked. *First in–first out (FIFO)* means that data is read from the first register in the stack. Conversely, *last in–first out (LIFO)* means that data is read from the last register in the stack. *Magnetic bubble memory* is a form of memory in which data is formed in bubbles, with the bubbles forming data loops.

Questions

1. What is the primary difference between the terms register and memory?

2. Give three examples of semiconductor memories.

3. What do the terms RAM, ROM, PROM, and RMM represent? Which ones are user-programmable?

4. What is the primary characteristic of content-addressable memory (CAM)?

5. What is the difference between serial data access and parallel data access?

6. What is a PLA? What are its three primary parts?

7. What is the difference between first in–first out (FIFO) and last in–first out (LIFO) memories?

8. What are the primary characteristics of magnetic bubble memory?

Problems

1. Given the programmable logic array (PLA) shown in FIG. 10-16, show how to create the outputs $X = A\overline{B} + \overline{A}\overline{B}C + C$, $Y = A\overline{B} + \overline{C}$, $Z = \overline{A}\overline{B}C$.

2. Use the PLA shown in FIG. 10-16 to create the outputs $X = \overline{A}B + A\overline{B}$, $Y = A\overline{B} + \overline{A}\overline{C}$, $Z = \overline{A}B + \overline{A}\overline{C}$.

3. In FIG. 10-11, the read/write select line is high to write and low to read. Show how this line should be connected to the read and write clocks in order to properly use the FIFO memory shown.

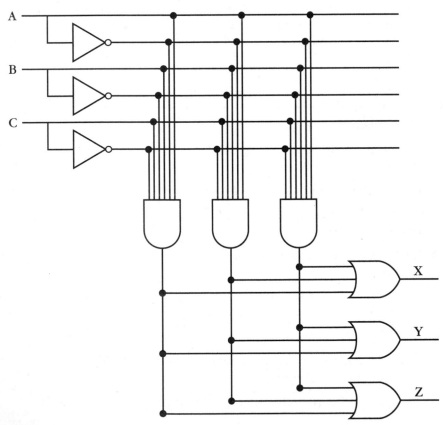

10-16

11

DIGITAL INTERFACING

AFTER YOU COMPLETE this chapter, you will be able to:

☐ Understand the different methods of transferring data
☐ Convert between digital and analog signals
☐ Explain the uses of the op amp in transmitting data

The dictionary definition of an interface is: *A common boundary between two parts of matter or space*. In the case of digital circuits, the "two parts of matter or space" can be interpreted to mean two physical or functional electronic units. Hence, an interface circuit could be one that transmits and receives digital data, say between two separate electronic chassis; or it might be one that converts analog information to digital data, or vice versa. This chapter explores some of the more common types of interface circuits likely to be encountered. These include buses, line drivers and receivers, digital-to-analog and analog-to-digital converters, concepts of series and parallel transfer, and the UART and RS-232C interface devices.

BUSES

Physically, a *bus* is a set of conductive paths that serves to interconnect two or more functional components of a system, or several diverse systems, together. Electrically, a bus is a collection of voltage levels and signals that allow the various devices connected to the bus to work properly together. For example, a microprocessor is connected to memories and input/output devices by certain bus structures. (See FIG. 11-1.) An *address bus* allows the microprocessor to address the memories, a *data bus* provides for data transfer between the microprocessor, the memory, and the input/output devices, and the *control bus* allows the microprocessor to control data flow and timing for the various components.

Buses can also be used to interconnect various instruments or other types of electronic systems. There are several widely used and accepted

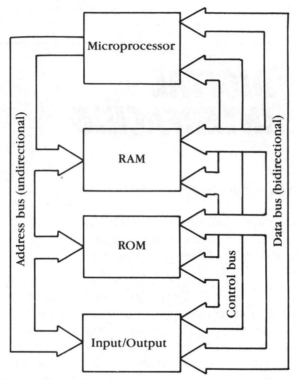

11-1   Buses within the microcomputer system. transfer all the data to and from the necessary components.

bus structures that have been established as so-called standards. The purpose of a standardized bus is to eliminate interface problems between various types of electronic equipment that conform to the bus specifications. If two pieces of equipment are designed to meet a given standard interface requirement, then when they are connected together via the bus, they will operate in conjunction with each other as intended without modification.

LINE DRIVERS AND RECEIVERS

Line drivers and receivers are the amplifier circuits that transfer digital data between units. Whenever such a data transfer occurs, problems of noise entering the system are also likely to occur. The noise may be ground noise between two chassis, or it may be crosstalk from signals inductively coupled between wires in cable harnesses. Whatever the cause, the basic solution to the noise problem lies in improving the signal-to-noise ratio at the receiving end. Three different approaches to improving signal-to-noise ratio will be considered.

Level Shifters

Probably the easiest way of improving signal-to-noise ratio is simply to increase the amplitude of the signal at the source. Given that the amount of noise on a particular transmission line remains constant, then the improvement in signal-to-noise ratio is in direct proportion to the increase in signal amplitude. Level shifters are the digital circuits that perform this function. A simple level shifter might consist of a common-emitter transistor amplifier, which receives signal at DTL or TTL logic levels and provides output voltage swings of $+ 12V$ and $- 12V$, to represent the logic 1 and logic 0 states, respectively. Disadvantages to this technique are the requirement for higher-voltage power supplies and the increased power consumption inherent in such a circuit.

Coaxial Cable Drivers

Another method for improving signal-to-noise ratio is by decreasing the amount of noise on the lines. This time, assuming that the *signal* amplitude remains constant, then the signal-to-noise ratio improvement is directly related to the reduction in *noise* at the receiving end. The most common method for noise reduction is through the use of properly terminated shielded cable, driven by low-impedance driver circuits. Here, coax drivers are the digital circuits that perform the digital data transmission. A straightforward example of a coax driver circuit is an emitter-follower transistor amplifier that drives a 50-ohm coax cable terminated at the receiving end with a 50-ohm resistor. The main disadvantage to this method is that shielded cable is quite expensive and system costs rise rapidly if large amounts of cable are required.

Differential Line Drivers and Receivers

Finally, a third method for improving signal-to-noise ratio is through differential amplifier techniques, utilizing the improvements gained through common-mode versus differential-mode gains. Differential line drivers and receivers are the digital circuits that perform this function. In general, if two wires are kept physically close to one another, as in a twisted pair, extraneous signals induced into one wire will also be induced into the other. Such signals, which occur simultaneously in two wires, are called common-mode signals. Similarly, any ground noise in two wires will also tend to be common-mode noise. If, on the other hand, the signals on the two wires swing in opposite directions, the signals are said to be differential-mode signals.

Differential-mode signals can be purposely generated, as shown in FIG. 11-2. The line driver that generates the differential mode signal is simply an AND gate and a NAND gate, with parallel inputs driving two sepa-

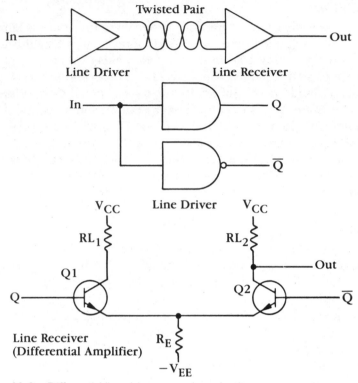

11-2   Differential line driver—receiver circuit.

rate wires. The input signal will appear with positive polarity on one wire and with inverted polarity on the other wire. The key to differential-mode operation is the line receiver circuit.

The line receiver consists of a basic differential amplifier circuit. To understand circuit operation, consider first the case where a positive signal is present at the base of Q1, while the base of Q2 is at ground potential. This is a typical differential-mode signal. Since the base of Q1 is positive, Q1 will conduct heavily, causing current to increase through resistor R_E. With the base of Q2 held at ground, this represents a reverse bias to transistor Q2, and current flow through R_E decreases by an amount equal to the increase caused by Q1. The result is that there is no net change in the current flow through R_E; thus, the emitters of both transistors can be considered as being at virtual ground. In this mode, the gain of the amplifier is quite high.

For common-mode operation, assume that the two bases are tied together, as is functionally the case if both base voltages change by the same amount and in the same direction. A positive voltage at the base of Q1 will cause an increase in current through R_E, but a similar increase at the base of Q2 also causes the same kind of current increase. To both transistors, the resultant positive voltage at their emitters is a negative feedback, causing a decrease in conduction through both transistors. Of course, this decrease in conduction offsets the increase caused by the base going more positive, thus very little signal gain is achieved.

What has been shown is that differential-mode signals (the true signals of interest) receive very high gain from the differential amplifier, while common-mode signals (noise) receive very low gain. The net result is a significant improvement in signal-to-noise ratio.

DIGITAL-TO-ANALOG CONVERTERS

Often, information that is in digital format is required in analog form. When such a conversion is needed, a digital-to-analog converter is used. An example of a requirement for such a conversion is the situation where a number of parallel digital data bits form audio or video information. In order to display the information on an oscilloscope or listen to the audio, the digital data must first be converted to an analog waveform. Typically, digital-to-analog conversion is performed with a resistive ladder network followed by an operational amplifier. Since each of these functions is somewhat unique, they are described separately.

Resistive Ladder Networks

An extremely simple method of converting binary digits to analog voltages recognizes the basic characteristics of binary data. There are only two voltage levels present for any binary number; thus, all bits of the number have the same voltage value when in their logic 1 states, and the same voltage value when in their logic 0 states. If a resistive network can be constructed which gives a large voltage output for the most significant bits of the number and smaller voltages for the least significant bits, then a digital-to-analog conversion will have been performed. Resistive networks which are scaled to perform this function are called *resistive ladders* or *ladder adders.* The latter name is derived from the fact that the network sums or adds the voltage contributions from the various bits to form the resultant analog voltage.

One very common type of resistive ladder is the binary ladder shown in FIG. 11-3. This ladder uses resistors which are selected in binary increments to form a voltage-divider network. Hence, the most significant bit uses a value of 1R, the next bit 2R, etc. The network is terminated with a resistor equal to the least-significant-bit resistor.

To understand the functioning of this ladder, assume that the logic levels happen to be $+5V$ for a logic 1 and ground for a logic 0. Also, assume that initially only the 2^5 bit is at a logic 1 and that all other bits are

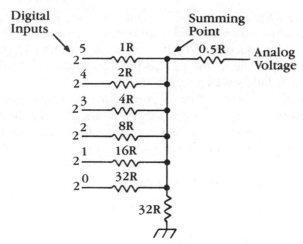

Basic Binary Ladder Network

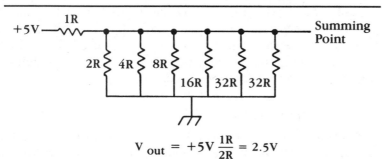

$$V_{out} = +5V\,\frac{1R}{2R} = 2.5V$$

(a) Equivalent Circuit for $2^5 = +5V$; all other bits = Ground

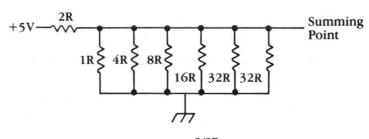

$$V_{out}\ +5V\,\frac{2/3R}{2\,2/3R} = 1.25V$$

(b) Equivalent Circuit for $2^4 = +5V$; all other bits = Ground

11-3   Binary resistive ladder network.

logic 0. A voltage divider is formed, consisting of one series resistor and the remaining resistors in parallel to ground. According to the rules for summing parallel resistances, it is seen that 32R in parallel with 32R becomes 16R, and 16R in parallel with 16R sums to 8R, and so on. The resultant total resistance to ground from the summing point is 1R. Obviously, if + 5V is divided between two resistors of value 1R, the point in the middle is at + 2.5V.

Now, assume that only the 2^4 bit is at a logic 1 and the rest go to logic 0. Here, the parallel resistors collapse to 2/3R, producing an output of 1.25V. This is one-half the voltage obtained when the 2^5 bit was a logic 1. From the above, it can be seen that each bit produces one-half the voltage of the next significant bit. The summing point becomes an analog voltage, proportional to the sum of all the individual voltages applied by the resistors.

One disadvantage to the binary ladder is that many different values of resistors are required, and also that each digital stage is terminated in a different resistance value. A scheme which uses uniform resistor values is the R − 2R ladder shown in FIG. 11-4.

Assume that the 2^5 bit is a logic 1 and that all other bits are logic 0. Starting at the bottom of the ladder, it is seen that 2R to ground in parallel with 2R to the 2^0 bit (also ground) sums to 1R. But 1R in series with 1R is 2R, and this is in parallel with 2R to ground for the 2^1 bit. Continuing the summing, the network collapses to a voltage divider consisting of two resistors of value 2R. Thus, when the 2^5 bit is a logic 1, the output of the ladder is + 2.5V, just as with the binary ladder. Considering the case where the 2^4 bit is a logic 1 and all other bits are logic 0, it is seen that the network collapses to a simple divider which this time produces an output of 1.25V. These results are identical with those obtained from the binary ladder network.

Operational Amplifiers

The resistive ladder networks previously described work well for driving very high impedance circuits. However, if an attempt is made to drive a low-impedance circuit, there is insufficient drive from the flip-flops that supply voltages to the individual resistors and a degradation of voltage levels occurs.

Uniform known voltage levels are a prerequisite for proper operation of the resistive ladder network. Operational amplifiers provide the necessary impedance matching and buffering as well as amplification of the analog voltage swings to achieve the well-behaved voltage levels.

Basically, an operational amplifier consists of a differential amplifier

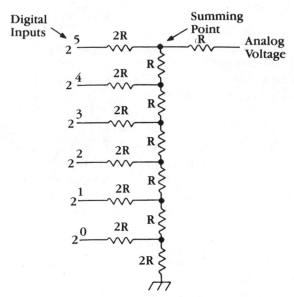

Basic R-2R Ladder Network

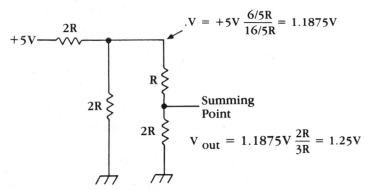

$$V_{out} \quad +5V\, \frac{2R}{4R} = 2.5V$$

A Equivalent circuit for $2^5 = +5v$; All other bits = ground.

$$.V = +5V\, \frac{6/5R}{16/5R} = 1.1875V$$

$$V_{out} = 1.1875V\, \frac{2R}{3R} = 1.25V$$

B Equivalent circuit for $2^4 = +5V$; All other bits = ground.

II-4   R-2R resistive ladder network.

quite similar to the one described previously (and shown in FIG. 11-2) followed by several stages of level shifting and additional value gain. The characteristics that distinguish an operational amplifier are: very high input impedance, very low output impedance, and a very large voltage gain. The high input impedance ensures that the amplifier won't load down the circuit driving it, while the low output impedance provides good drive capability to other circuits. By having a large voltage gain, negative feedback can be employed (usually in the form of a feedback resistance, R_F) to control the overall gain of the amplifier in a particular application.

Some of the most common operational-amplifier configurations are shown in FIG. 11-5. Note that each configuration has a plus input and a minus input, corresponding to the two inputs of a differential amplifier.

Inverting Amplifier   Probably the most-used configuration is the standard inverting amplifier shown at the top of the figure. Feedback resistor R_F, in conjunction with the input resistance, R_i, forms the negative feedback path which determines the gain of the circuit. For this configuration, the voltage gain is defined as the ratio R_F/R_i. Thus, for example, when $R_F = R_i$, then the gain of the circuit is unity. Similarly, if $R_F = 10k$ and $R_i = 5k$, the gain of the circuit is two. In other words, for an output signal amplitude of 2V, an output amplitude of 4V will be obtained.

Noninverting Amplifier   The noninverting amplifier configuration is quite similar to the inverting amplifier, except that it uses the other input to the differential amplifier. The gain of this circuit is $1 + R_F/R_i$. A very common setup here is the special case where a unity-gain voltage follower is desired. From the equation, it is seen that in order to obtain unity gain, R_F should equal zero and R_i should equl infinity. This is equivalent to saying that R_i does not exist and R_F is a short circuit. Under these conditions, the circuit shown in FIG. 11-6 results. The unity-gain voltage follower is often used where there is no change in signal level desired, but buffering is required. As previously noted, the operational amplifier's high input impedance does not load the input circuit, while the low output impedance provides excellent drive capability.

Summing and Difference Amplifiers   Referring again to FIG. 11-5, the summing amplifier configuration is seen to be a special case of the inverting amplifier. Of particular interest here is the fact that the summing amplifier input circuit looks very much like the two ladder adder networks previously described. Indeed, when the ladder adder network and operational amplifier are combined, the resulting configuration is that of a digital-to-analog converter. A diagram of a complete digital-to-analog converter is shown in FIG. 11-7. Note that the operational amplifier has a ladder adder circuit at its input. Feedback is provided by a single resistor, which

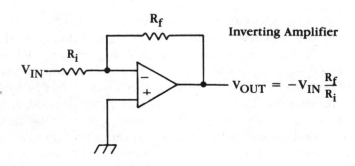

Inverting Amplifier

$$V_{OUT} = -V_{IN}\,\frac{R_f}{R_i}$$

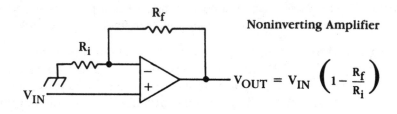

Noninverting Amplifier

$$V_{OUT} = V_{IN}\left(1 - \frac{R_f}{R_i}\right)$$

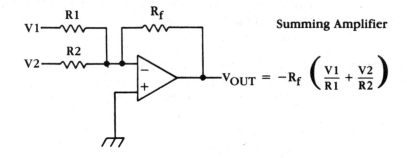

Summing Amplifier

$$V_{OUT} = -R_f\left(\frac{V1}{R1} + \frac{V2}{R2}\right)$$

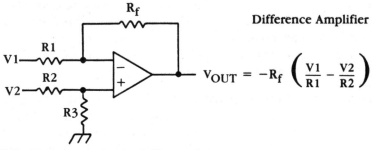

Difference Amplifier

$$V_{OUT} = -R_f\left(\frac{V1}{R1} - \frac{V2}{R2}\right)$$

11-5 Typical operational amplifier configurations.

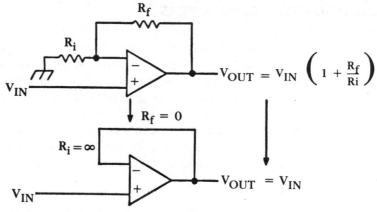

11-6 Noninverting unity-gain amplifier.

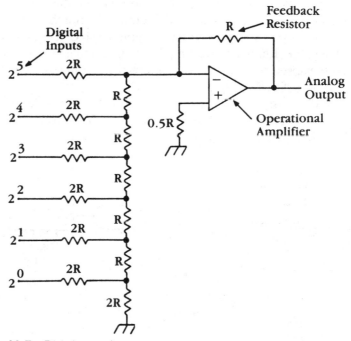

11-7 Digital-to-analog converter.

in this case gives unity gain. The main function of the operational amplifier in this circuit is to provide the drive necessary for the following circuits and to buffer the resistive ladder (and flip-flops driving it) from the effects of circuit loading. The resistor on the plus input of the amplifier is included so that the differential amplifier sees approximately the same impedance on both inputs.

ANALOG-TO-DIGITAL CONVERTERS

Analog information exists in many different forms. This information may be represented by a continuously variable voltage, or it may be represented by the position of a shaft or lever. Also, the analog action can range from quite slow, as with a shaft rotation of several revolutions per minute, to very fast, as is the case of a waveform varying at a rate of several megahertz. Because of the wide variety of analog devices, there is also a corresponding variety of analog-to-digital conversion techniques. The following paragraphs will not attempt to explore all of the known techniques, but rather will describe the basics of several commonly used analog-to-digital converters.

Elapsed-Time Technique

One simple way of converting an analog voltage to a digital number is by generating an analog ramp voltage and counting the number of clock periods it takes the ramp voltage to rise to a value which equals an unknown analog voltage. Providing that the ramp is linear and that its slope remains constant, the elapsed time to reach a given voltage is directly proportional to the amplitude of that voltage.

A block diagram of a converter using this technique is shown in FIG. 11-8. The top portion of the figure represents the basic converter, which will be described first.

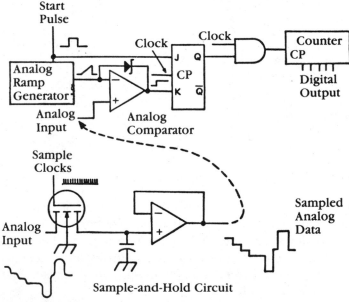

11-8   Elapsed time technique of analog-to-digital converter.

The analog ramp generates a sawtooth waveform, starting at the most negative voltage to be encountered and graduating to the most positive voltage expected. The input voltage is compared in an analog comparator against the ramp voltage. The analog comparator is simply an operational amplifier with a zener diode for feedback.

If the ramp voltage is greater than the analog input, the comparator output equals the zener voltage. If the ramp amplitude is less than the analog input, the comparator output is essentially at ground potential. The flip-flop acts as a control for the counter. Upon receipt of a start pulse, the ramp starts at its most negative value, the flip-flop is set, and clock pulses are applied to the counter. As the ramp voltage increases, so does the counter. When the ramp reaches the same value as the input voltage, the comparator output goes positive, resetting the flip-flop and stopping the clock pulses to the counter. At this time, the state of the counter represents the analog input voltage.

If the analog voltage changes very slowly, the above circuit is sufficient. However, with a more rapidly changing analog input, the voltage may change significantly between the time the ramp is started and the time when the conversion is completed. A method for holding the analog input steady is the sample-and-hold circuit shown at the bottom of FIG. 11-8.

Here, a MOS transmission gate is used to charge a capacitor connected to an operational amplifier. Each time a clock pulse occurs, the capacitor charges. Between clock pulses, the MOS gate is open and the capacitor stores a constant voltage until the next clock pulse occurs. In this way, the analog-to-digital converter is presented with a constant voltage for one entire conversion cycle; it is presented with some other constant voltage for the next conversion cycle.

Binary Ramp Comparison Technique

One problem with the elapsed-time technique is that the generated ramp must be very smooth and linear if accurate results are to be obtained. An alternative approach is to generate a ramp by clocking a digital counter and performing a digital-to-analog conversion on the counter outputs. Now, if the ramp is compared to the analog input, the point at which the ramp equals the input voltage can again be used to stop the counter. The counter state represents the value of the analog input voltage.

The primary difference is that the generation of the ramp is not dependent on time. If desired, the ramp could be increased by a small amount, stopped, then increased some more. In either case, the conversion would be identical to that obtained with a smoothly increasing ramp. Of course, it is almost always advantageous to complete a conversion cycle

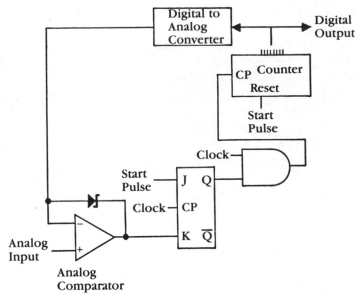

11-9 Binary ramp comparison technique.

as quickly as possible, mainly because the analog input is liable to change over a period of time.

A block diagram of a binary ramp converter is shown in FIG. 11-9. The conversion cycle starts by resetting the counter to all zeroes and setting the clock control flip-flop, thus applying clock pulses to the counter. As the count increases, the digital-to-analog converter changes the output states of the counter to an analog ramp voltage. When the ramp voltage equals the analog input, the flip-flop is reset and the conversion cycle is complete. At this time, the state of the counter represents the magnitude of the analog input. As with the previous circuit, a sample-and-hold circuit can be used if needed to hold the analog input steady during the conversion cycle.

Successive Approximation Technique

The binary ramp technique has one major drawback. If the maximum possible positive voltage were to be converted in a 6-bit converter, the conversion process would require $2^6 = 64$ steps to complete—one for each possible count of the binary counter. Thus, the binary ramp is extremely slow. A method that greatly reduces the conversion time is the successive approximation technique. As implied by its name, this method successively approximates the input voltage by factors of two, to very rapidly converge to the correct answer. As will be seen, instead of 64 steps, the successive

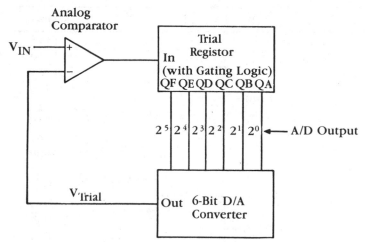

11-10   Block diagram of successive approximation converter.

approximation technique requires only 6 steps to perform the same 6-bit conversion.

A block diagram of a 6-bit successive approximation converter is shown in FIG. 11-10. Initially, the assumption is made that the number representing the input analog voltage is 100000. Note that this value is exactly halfway between 0V and the maximum possible voltage. The number 100000 is loaded into the trial register and converted to an analog voltage. At this time, the analog comparator has two inputs, the input voltage (V_{in}) and the trial voltage (V_t).

If the input voltage is greater than the analog V_t representation of 100000, the comparator output is a logic 1 and the most significant bit of the trial register remains a 1.

If the input voltage is less than the trial voltage, the comparator output is a logic 0 and the most significant bit is set to 0. For the second step, the next bit in line is set to a logic 1, while retaining the previously determined most significant bit in the trial register. This has the effect of picking a new trial voltage halfway between the remaining limits to be determined, either 110000 or 010000.

Again a comparison is made to see in which direction the input voltage lies with respect to the trial voltage. Again, if the voltage is more positive, the comparator output is a logic 1, indicating that a 1 should be in that bit position. If the input voltage is more negative than the trial voltage, the comparator output is a logic 0 and a 0 would be placed in the bit position that is now most significant. This process is repeated six times, each time halving the remaining voltage to arrive at the correct voltage more quickly, and each time writing a 1 or 0 in the appropriate bit posi-

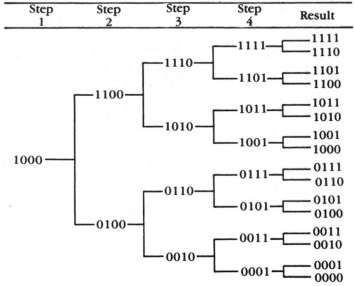

Step 1	Step 2	Step 3	Step 4	Result

11-11   Successive approximation trial steps.

tion, based on the results of the comparison between the new trial voltage and the input voltage. Thus, the successive approximation technique performs a conversion in n steps (where n is the number of bits in the number) instead of in 2^n steps, which may be required with the binary ramp technique.

FIGURE 11-11 illustrates the steps which would be followed for the 4-bit conversion. At each step, 1 of 2 possible branches is picked for a new trial number, then a comparison is made to determine the actual number. The process can be extended to 5, 6, or even 10 bits, if desired. The main limiting factor is the small differential voltage which will be applied to the comparator as the trial voltage approaches the input voltage.

Optical Shaft Encoder

The previous techniques of analog-to-digital conversion use electronic devices entirely. However, many analog events are mechanical actions. In this case, an electromechanical device must be used to perform the conversion. A very frequently used electromechanical analog-to-digital converter is an optical shaft encoder. A simplified diagram of an encoder of this type is shown in FIG. 11-12. The encoder consists of a code wheel with opaque and transparent areas, mounted on a shaft. A light source shines through the code wheel onto three photocells, each of which emits a voltage proportional to its light inputs.

If no light shines on a photocell, its output is low, representing a logic 0. If the light does shine on a photocell, that cell's output is high, repre-

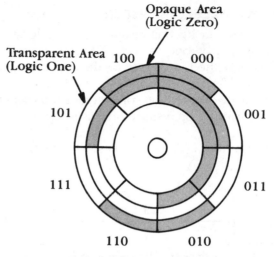

Code Wheel (Three-Bit Gray Code)

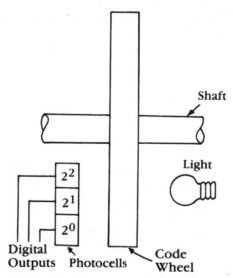

11-12 Optical shaft encoder.

senting a logic 1. By coding the wheel such that there are eight discrete outputs for an infinite variety of shaft positions, an analog-to-digital conversion has been performed. The particular code wheel in this example is constructed so that its outputs are gray-code representations of the shaft positions. Thus, if the wheel happens to be at some position between two different code outputs, the worst error which will be obtained is a single bit.

SERIAL AND PARALLEL TRANSFER

Data in binary form can be transferred from one location to another within a digital system by one or both of two basic methods: *serial* or *parallel*. These two methods are based on the relationship of bits as they are being transferred from place to place.

Serial data means that the bits follow one another so that only one bit at a time is transferred on a single line, as is shown in FIG. 11-13. The

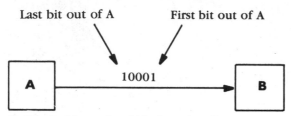

11-13   Serial transfer of bits from A to B.

rate at which the bits are transferred from A to B is called the *baud rate* and is expressed in bits per second. For instance, if 300 bits are transferred from A to B in one second, the rate is 300 baud.

Parallel data means that all the bits in a given group are transferred simultaneously on separate lines, as illustrated in FIG. 11-14.

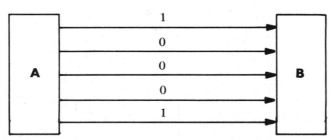

11-14   Parallel transfer of bits from A to B. All bits leave A at the same time.

UART

The *u*niversal *a*synchronous *r*eceiver *t*ransmitter, or UART, is an interfacing device that converts serial data to parallel data and vice versa. This feature is useful when, for example, a microprocessor-based system must communicate with external devices that send and/or receive serial data (recall that a microprocessor sends and receives parallel bits on a data bus). FIG-URE 11-15 illustrates the UART in a microprocessor-based system.

The UART receives data in serial form, converts it to parallel form, and places it on the data bus. The UART also accepts parallel data from the data

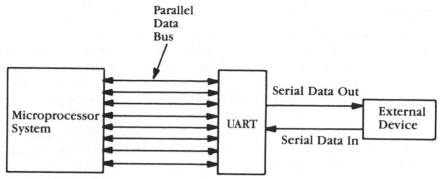

11-15 UART interface unit.

bus, converts it to serial form and transmits it to an external device. Typical examples of devices that operate with serial data are teletypes and certain communication systems. FIGURE 11–16 illustrates these basic conversions for an eight-bit data bus.

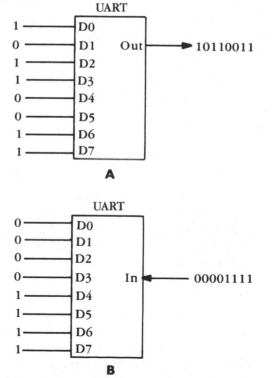

11-16 Basic operation of UART. (A) shows parallel-to-serial conversion (transmission) and (B) shows serial-to-parallel conversion (receiving).

RS-232C

The most widely used interface device for serial data communications is the RS-232C standard as identified by the Electronic Industries Association (EIA). This standard is used for interface between data terminal equipment and data communication equipment. Interfacing a computer and a peripheral device, such as a printer or CRT, would be one application.

Basically, RS-232C specifies a 25-pin connector and assigns serial signals to specific pins on the connector. Data rates up to 20 kilobaud can be accommodated under this specification. Four types of data bytes are defined: data signals, control signals, timing signals, and grounds.

A logic 1 signal is defined as a voltage between $-3V$ and -25 V. A logic 0 is between $+3V$ and $+25V$. Control signals are considered to be off if in the negative range (a logic 1 in this case) and they are considered on in the positive region (a logic 0). There can be no voltages between $-3V$ and $+3V$. Special IC devices are available to translate from RS-232C levels to TTL and CMOS levels and vice versa.

Of the 25 signal lines defined by the RS-232C standard, 2 are grounds, 4 are data signals, 12 are control signals, and 3 are timing signals. Each signal has a particular nomenclature, abbreviation, and pin assignment. In any given application, a piece of equipment does not have to use all signals provided for. The number of specified signals actually used varies from one application to another. This standard, as well as the other standards used in the industry for interfacing, exist to facilitate matching the equipment that conforms to the standard.

Summary

In order for information to be useful, it must get from one place to another. An electronic interface circuit is used to transmit data between two different devices or between two sections of the same device. This is normally done using buses, which carry the data. In a microprocessor, there are three primary buses: address, data, and control. In order to avoid noise problems on the buses, several different techniques are used, among them, level shifters, coaxial cable drivers, and differential drivers and receivers. Real-world data is normally in analog form, so when data is transmitted to or from the real world, digital-to-analog or analog-to-digital converters are used. Operational amplifiers, which are high-impedance differential amplifiers, are used in both operations.

Data is transferred in either parallel or serial form. Parallel transfer has the advantage of speed but requires additional hardware. A universal asynchronous receiver transmitter (UART) is used to convert parallel data to or from serial form, for transmission between devices. The most common UART for serial data transmission is the RS-232C interface.

Questions

1. In electronics, what is meant by the term *interface?* How is interfacing usually achieved?

2. What is a *bus?* What are three different buses used in microprocessors? What does each do?

3. What are *line drivers? Receivers?* What is their function in interface circuits?

4. What are three ways to reduce the effects of noise in an interface circuit?

5. What is the function of an analog-to-digital converter? Why is it used?

6. What is an *operational amplifier?*

7. List four different configurations using operational amplifiers and sketch their basic circuitry.

8. What is the function of an analog-to-digital converter? Why is it used?

9. What are three means by which analog-to-digital conversion may occur electronically?

10. What is the difference between serial and parallel data transfer? What advantage does each have?

11. What does *UART* stand for? What is the function of the RS-232C interface?

Problems

1. An 8-bit digital signal is to be converted to an analog signal, as in FIG. 11-3. If the resistor between the summing point and the output is 1 kΩ,   **a.** draw the circuit   **b.** calculate the value of each of the resistors

2. Repeat Problem 1 using an R-2R ladder network, as in FIG. 11-4.

3. Assuming that the op amp does not go to saturation, calculate the output voltage for each of the circuits in FIG. 11-17.

4. Repeat Problem 3 for each of the circuits in FIG. 11-18.

5. A 6-bit digital word is to be used to represent voltages from 0 to 5 V. Calculate the voltage value for each bit.

6. An 8-bit digital word is to be used to represent voltages from 0 to 12 V. Calculate the voltage value for each bit.

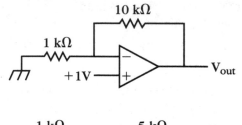

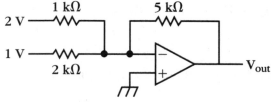

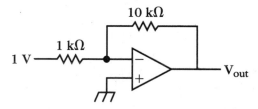

11-17

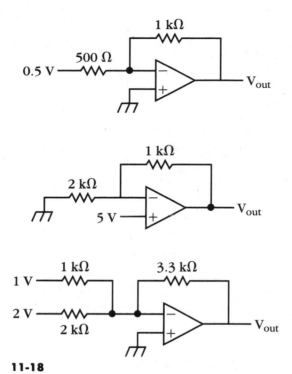

11-18

DIGITAL DISPLAYS

AFTER YOU COMPLETE this chapter, you will be able to:

☐ Describe the different types of individual lamps used in digital displays
☐ Discuss the primary characteristics of LED lamps
☐ Understand the different methods with which segments are used to create displays

There are many occasions when digital data from various portions of a system are displayed. Sometimes the data is displayed in raw binary format, but more often it is converted to some type of decimal display so that operators can more readily interpret the results being displayed. It is not unusual for a digital display to represent the end result of an entire set of computations and logic operations. Some examples of equipments where this is the case are counters, digital voltmeters, and electronic calculators. Because digital displays are so frequently used, it is desirable to gain some familiarity with the different types of displays available. This chapter describes the most common display types encountered.

TYPES OF INDIVIDUAL LAMPS

Almost everyone is familiar with the little red light associated with the power switch on a piece of test equipment. This light is the simplest form of a digital display. When the light is lit, the power is on (a logic 1), and when the light is dark, the power is off (a logic 0). There is no universal lamp type which is used everywhere. The type of lamp used is a function of the particular application. Some lamp types found in digital systems are incandescent, neon, fluorescent, and light-emitting diodes.

Incandescent

An example of an incandescent lamp is the standard light bulb used in the home. These lamps operate by sending current through a very fine filament wire, causing the wire to heat up and thus emit light. Although alternating current is normally used in the home, the same principle applies if direct current is used. In general, the incandescent lamps made for digital equipment are small and can operate on low voltages. The main faults with all incandescent lamps is that their filaments tend to burn out fairly often and they waste energy because they generate heat.

Fluorescent

An incandescent lamp gives off light due to heating of its filament. Correspondingly, it is known that an ordinary vacuum tube uses the heat from a filament in conjunction with a fairly large anode voltage to cause thermionic emission of electrons from the cathode that are attracted to the anode. Fluorescent lamps of the type used in digital equipment utilize a very similar technique. The fluorescent lamp is basically a vacuum tube diode in which the anode has been coated with a special phosphor that glows when bombarded with electrons. When the diode conducts, the lamp is on; and when the diode is off, the lamp is off.

Several fluorescent lamps are shown in FIG. 12–1. The basic lamp sche-

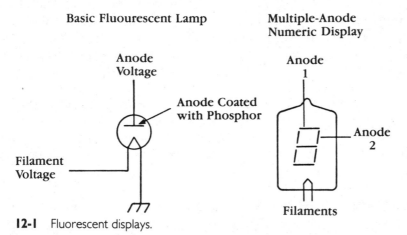

12-1   Fluorescent displays.

matic simply shows the expected diode structure. This basic lamp configuration is not frequently found. The more common fluorescent display is the numeric display. Here, the numbers are formed by placing many anode segments in a single envelope, then applying separate anode voltages to each as necessary to turn on segments which form the desired number. Typically, fluorescent displays are greenish-blue in color.

Neon

In direct contrast to the incandescent and fluorescent lamps which utilize hot cathodes for thermionic emission, neon lamps are often referred to as cold-cathode or gas discharge displays.

The reason for this is in their basic functional mechanism: A neon bulb consists of a glass envelope filled with neon gas and containing an anode and a cathode. There is no filament. When a high voltage is applied between the cathode and anode terminals, the neon gas in the tube ionizes, thus causing an orange glow. Normally, neon lamps are quite tiny and are well suited for behind-the-panel displays. One typical use of this type is shown in FIG. 12–2. Seven individual neon lamps are placed behind a

12-2   Mask for seven neon lamps.

mask consisting of an opaque material with slots cut out where the lamps can show through. By turning on various combinations of neon bulbs, any number from 0 through 9 can be formed. One disadvantage to neon lamps is the high ionization voltage required.

Light-Emitting Diodes

A light-emitting diode (LED) is simply a semiconductor diode that is made from a material such as gallium phosphide or gallium arsenide phosphide instead of the usual silicon or germanium normally employed for diodes. As might be guessed from the phosphide name of the material, this material emits light when properly stimulated. Actually, almost any *pn* junction emits visible light, but the efficiency of the radiation to input energy is typically quite low; also, the concentration of radiated energy may sometimes fall outside the visible light spectrum.

For reference purposes, FIG. 12–3 shows the visible light range in terms of the response curve for the human eye to various wavelengths of radiation. Notice that the color green is near the peak of the curve, while red is much harder for the eye to detect. Hence, it would seem that all light-emitting diodes would emit green light. In reality, the efficiency of green emitting diodes is so low compared to red diodes that the great majority of all light-emitting diodes are red.

The physical basis for radiation in a semiconductor diode is the recombination of holes and electrons that occurs in the depletion region

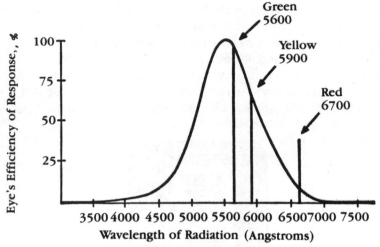

12-3   Eye's response to radiation.

and at the *pn* junction whenever a diode is forward biased. Referring to FIG. 12–4, it is seen that when a diode is biased such that the *p* side of the diode is more positive than the *n* side, the diode is considered forward biased and a large current will flow. This is basic to the operation of all

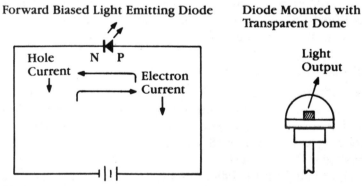

12-4   Light-emitting diode mechanism.

semiconductor diodes. Further, the resulting current flow can be considered to consist of electrons traveling from the negative battery terminal to the positive terminal, and concurrently, it can be considered as a flow of holes from the positive terminal to the negative battery terminal. For the moment, consider only electron current flow.

The large concentration of electrons flowing through the junction is going to result in an excess of electrons also flowing through the depleted *n* material in the immediate area of the junction. Some of the electrons will therefore tend to recombine with holes in the depletion region of the *n* material in an attempt to overcome the space-charge effect in that area.

Similarly, since a hole current of equal magnitude must flow in the opposite direction, the same type of recombination will occur in the depletion region at the *p* side of the junction. From physics, it is known that when recombination does occur, photon emission (the generation of light) also occurs. This emission or radiation is the output that is seen from a light-emitting diode. Through variation of the ratios of gallium arsenide and gallium phosphide in the *p* and *n* materials, the wavelength of the emission can be made to vary from approximately 5500 to 9000 angstroms. From FIG. 12-3, this is seen to range from green light to well beyond the visible red light spectrum. Other materials can be chosen to emit radiation far into the infrared portion of the spectrum if desired.

DISPLAYS

The types of displays discussed in the remainder of this chapter include grouped binary, multiplane, seven-segment, LED arrays, scanned alphanumeric, and liquid crystal displays.

Grouped Binary Displays

A long row of lights displaying binary information can be quite a confusing sight, and the pattern of OFF and ON lights is virtually impossible to remember. Yet, some means must be provided for interpreting raw binary data. The method normally used is an octal or BCD grouping, as shown in FIG. 12-5. If the data is in straight binary format, a grouping of each three bits allows reading octal numbers directly from the binary. Note that in the example, 452_8 is much easier to remember than 100101010_2. If the data happens to be in BCD format, then 4-bit groupings provide a natural readout for each of the BCD decades.

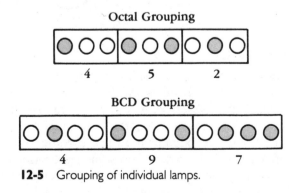

12-5 Grouping of individual lamps.

Multiplane Displays

One of the first digital displays to provide data in a readable decimal format was the multiplane neon tube. This tube is a neon-filled gas envelope, with a single anode and 10 different cathodes. Each cathode is shaped to form one decimal digit, with the digits stacked one behind another. The plane which is ionized glows, thereby displaying the digit formed by the cathode. A multiplane display is shown in FIG. 12-6. The BCD-to-decimal decoder assures that only one plane will be energized at any given time.

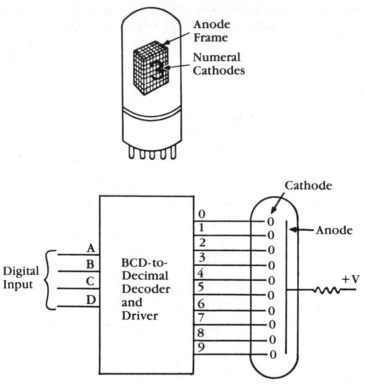

12-6   Multiplane neon display.

Seven-Segment Displays

An improvement over multiplane displays is the seven-segment display shown in FIG. 12-7. This display shows any desired digit on a single plane. The display is made up of seven segments, which can be illuminated in various combinations to form any one of the 10 decimal digits. Usually, the seven-segment codes are decoded from BCD according to the truth table shown. There may be a slight variation in some applications, depending on whether the six and nine digits include a tail. For example, if a tail were included on the digit six, the "a" segment (top horizontal bar) would also

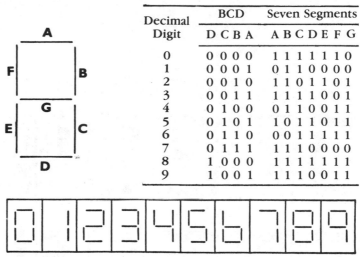

| Decimal | BCD | Seven Segments |
Digit	D C B A	A B C D E F G
0	0 0 0 0	1 1 1 1 1 1 0
1	0 0 0 1	0 1 1 0 0 0 0
2	0 0 1 0	1 1 0 1 1 0 1
3	0 0 1 1	1 1 1 1 0 0 1
4	0 1 0 0	0 1 1 0 0 1 1
5	0 1 0 1	1 0 1 1 0 1 1
6	0 1 1 0	0 0 1 1 1 1 1
7	0 1 1 1	1 1 1 0 0 0 0
8	1 0 0 0	1 1 1 1 1 1 1
9	1 0 0 1	1 1 1 0 0 1 1

12-7   Seven-segment display.

be illuminated. Seven-segment displays are found using any of the previously mentioned lamp types.

Light-Emitting-Diode Arrays

The seven-segment display is quite good, but the decimal digits are slightly crude; and if one segment happens to burn out, an incorrect display can result. For example, if the G segment fails, an 8 will be displayed as a 0. An even more realistic set of numerals is possible by using an array of LEDs formed into a matrix consisting of several rows and columns. A typical 5-×-7 matrix is shown in FIG. 12-8. This matrix has a noticeable advantage: not only do the numerals appear better formed, but if one diode in the array should fail, the displayed number is still quite readable. The entire array is usually contained in the same integrated-circuit package as its decoder. Therefore, an LED array is normally addressed with a 4-bit BCD code and no external decoding is necessary.

Scanned Alphanumeric Displays

The array concept can easily be expanded to include display of letters as well as numbers. Such a readout is called an *alphanumeric display*. Since a simple BCD decoder will not suffice for the large quantity of characters possible, this type of array is typically driven by a read-only memory addressed by a binary code. The binary code must contain 10 combinations for the 10 decimal digits, 26 combinations for the letters of the alphabet, and enough other combinations to display such special characters as pe-

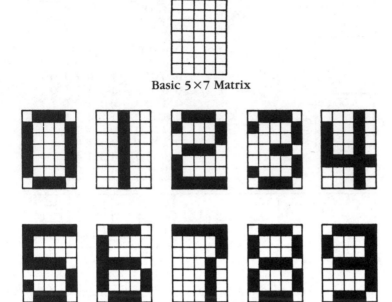

Basic 5×7 Matrix

12-8   Light-emitting-diode array.

riods, question marks, exclamation points, colons, and the like. Since at least 36 combinations are required, the smallest number of binary bits which can be used is 6. One commonly used 6-bit alphanumeric code is called ASCII. Table 12-1 shows the ASCII code combinations and the characters they represent. The function of the read-only memory is to convert the binary code bits to the proper combinations necessary to drive the 5- × -7 array.

Any meaningful alphanumeric display will consist of several characters displayed simultaneously. Each character displayed requires a separate 5- × -7 array. Under these conditions, it becomes feasible to drive the arrays by a method known as *scanning,* or *strobing.* This method consists of applying a pulse to one row of all the displays, and simultaneously applying data bits to the individual columns. Each row is sequentially strobed (pulsed). If the strobing occurs at a fairly high rate, the displays do not appear to flicker.

A vertical strobing arrangement for five characters is shown in FIG. 12-9. With the strobing technique, the data bits from the read-only memories must change each time the strobe changes. In this way, the input to a given column appears as a 7-bit serial data stream, which is continuously recirculated. Because the data stream changes in synchronism with the strobe lines, each row of the array always sees the same bit combination.

Table 12-1.   Six-Bit ASCII Code.

A6	A5	A4	A3	A2	A1	Character	A6	A5	A4	A3	A2	A1	Character
0	0	0	0	0	0	@	1	0	0	0	0	0	Blank
0	0	0	0	0	1	A	1	0	0	0	0	1	!
0	0	0	0	1	0	B	1	0	0	0	1	0	"
0	0	0	0	1	1	C	1	0	0	0	1	1	#
0	0	0	1	0	0	D	1	0	0	1	0	0	$
0	0	0	1	0	1	E	1	0	0	1	0	1	%
0	0	0	1	1	0	F	1	0	0	1	1	0	&
0	0	0	1	1	1	G	1	0	0	1	1	1	'
0	0	1	0	0	0	H	1	0	1	0	0	0	(
0	0	1	0	0	1	I	1	0	1	0	0	1	)
0	0	1	0	1	0	J	1	0	1	0	1	0	*
0	0	1	0	1	1	K	1	0	1	0	1	1	+
0	0	1	1	0	0	L	1	0	1	1	0	0	,
0	0	1	1	0	1	M	1	0	1	1	0	1	-
0	0	1	1	1	0	N	1	0	1	1	1	0	.
0	0	1	1	1	1	O	1	0	1	1	1	1	/
0	1	0	0	0	0	P	1	1	0	0	0	0	0
0	1	0	0	0	1	Q	1	1	0	0	0	1	1
0	1	0	0	1	0	R	1	1	0	0	1	0	2
0	1	0	0	1	1	S	1	1	0	0	1	1	3
0	1	0	1	0	0	T	1	1	0	1	0	0	4
0	1	0	1	0	1	U	1	1	0	1	0	1	5
0	1	0	1	1	0	V	1	1	0	1	1	0	6
0	1	0	1	1	1	W	1	1	0	1	1	1	7
0	1	1	0	0	0	X	1	1	1	0	0	0	8
0	1	1	0	0	1	Y	1	1	1	0	0	1	9
0	1	1	0	1	0	Z	1	1	1	0	1	0	:
0	1	1	0	1	1	[	1	1	1	0	1	1	;
0	1	1	1	0	0	\	1	1	1	1	0	0	<
0	1	1	1	0	1	]	1	1	1	1	0	1	=
0	1	1	1	1	0	^	1	1	1	1	1	0	>
0	1	1	1	1	1	—	1	1	1	1	1	1	?

Liquid Crystal Display

Liquid crystal displays (LCDs) are usually arranged in the same seven-segment numerical format as the LED display. There are two types of LCD: the *dynamic-scattering type* and the *field-effect type.* The construction of a dynamic-scattering type of liquid crystal cell is illustrated in FIG. 12-10. The liquid crystal material can be one of several organic compounds that exhibit the optical properties of a crystal although they remain in liquid

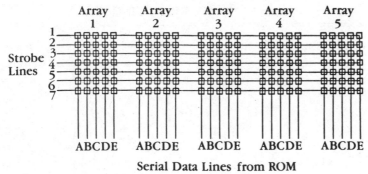

Serial Data Lines from ROM

12-9   Vertically strobed alphanumeric display.

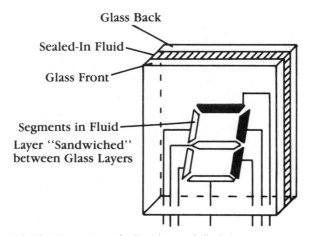

12-10   Illustration of a liquid crystal display.

form. Liquid crystal is layered between glass sheets with transparent electrodes deposited on the inside faces. When a potential is applied across the cell, charge carriers flowing through the liquid disrupt the molecular alignment and produce turbulence. When not activated, liquid crystal is transparent. When activated, the molecular turbulence causes light to scatter in all directions; the phenomenon is called *dynamic scattering.*

The construction of a field-effect LCD is similar to that of the dynamic-scattering type except that two thin polarizing optical filters are located inside each glass sheet. The liquid crystal material in the field-effect type is called *twisted nematic.* This type of cell material actually twists the light passing through it when the cell is not energized, which allows light to pass through the optical filters, and the cell appears bright (it can also be made to appear dark). When the cell is energized, the light does not become twisted and the cell remains dull.

Liquid crystal cells can be transmittive or reflective. In the *transmittive* cell, both glass sheets are transparent so light from a rear source scatters forward (toward the viewer) when activated. The *reflective* type has a reflecting surface on one of the glass sheets. In this case, incident light on the front surface of the cell is dynamically scattered when activated. Both the transmittive and reflective types of cells appear quite bright even under high-ambient-light conditions.

Since liquid crystal cells are light reflectors or transmitters rather than light generators, they consume very little energy. The only energy required is that needed to activate the liquid crystal. The total current flow through four, small, seven-segment LCD displays is about 25μA for dynamic-scattering cells and 300μA for field-effect cells. However, the LCD requires an ac voltage supply, either in the form of a sine wave or a square wave. The reason is because a direct current produces a plating of cell electrodes, which could damage the device. A typical supply for a dynamic-scattering LCD is 230V peak-to-peak square wave with a frequency of 60 Hz. A field-effect cell typically uses 8V peak-to-peak.

Summary

Displays are a primary form of digital output. Once a result has been determined, it is necessary to display the output. Digital output may be in the form of individual lamps, each representing a single bit, or a display format, where the output is typically in numeric or alphanumeric form. Individual displays are typically incandescent, fluorescent, neon, or light-emitting diodes (LEDs).

Grouped displays may use the same materials as individual displays. The more common use an LED or liquid crystal display (LCD) array, either in the form of seven segments or as a matrix. In a seven-segment display, binary or BCD information is converted to inputs which turn on individual segments. In a matrix, the display is considered a grid, and individual locations within the grid are energized to display alphanumeric characters.

Questions

1. Identify the four primary types of individual lamp displays.

2. For each of the individual lamp displays, describe   **a.** the relative power usage   **b.** the brightness   **c.** the color

3. What is a *grouped binary display?*

4. What is a *seven-segment display?* What are its advantages and disadvantages?

5. What is a *matrix display?* What are its advantages and disadvantages?

6. What does *LCD* mean? What are the two different types of LCDs?

Problems

1. A BCD input is to be connected to a seven-segment display, as in FIG. 12-7. Write the logic expression for segments *A, B, C,* and *D.* Using Boolean algebra and/or Karnaugh maps, simplify your answers.

2. Repeat Problem 1 for segments *E, F,* and *G.*

THE MICROPROCESSOR

AFTER YOU COMPLETE this chapter, you will be able to:

☐ Describe the different elements and the function of a micropro-
cessor
☐ Describe the different types of memory operations
☐ Explain the meaning of the term instruction set
☐ Understand input/output operations

This chapter provides an introduction to the microprocessor, a digital device that has had a major impact on the electronics industry and is the foundation of microcomputer technology. The purpose of this coverage is to give you a fundamental understanding of what
a microprocessor is and how it functions.

A microprocessor is a large-scale integrated circuit (LSI), which
means it contains at least 100 equivalent gates per IC (chip). It contains the
processing portion of a microcomputer. Physically, a microprocessor
comes in a single component as in FIG. 13-1. A symbol commonly used to
stand for microprocessor is μP (the Greek letter *mu* represents the prefix
micro).

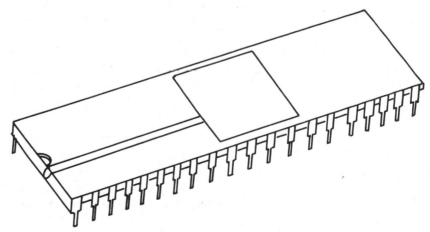

13-1 Microprocessor package.

THE MICROCOMPUTER

Because a microprocessor is part of a computer, let's first get a basic idea of what a microcomputer is, and distinguish it from a microprocessor. Figure 13-2 shows a block diagram of a basic microcomputer. It consists of

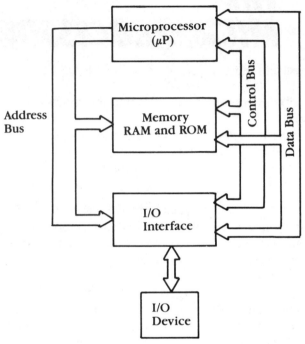

13-2   Microcomputer block diagram.

four blocks: the microprocessor, memory, input/output (I/O) interface, and I/O device. The microprocessor unit within the microcomputer is interconnected to the memory and I/O interface via an address bus and a data bus.

The function of the memory is to store binary data that is to be used or processed by the microprocessor. The function of the I/O interface is to get information in and out of the memory or the microprocessor from the I/O device. The I/O device could be a keyboard, video terminal, printer, magnetic disk, or other type of equipment.

The address bus provides a path from the microprocessor to memory and I/O interface. It allows the microprocessor to select the memory address from which data is stored. It also provides for communication with an I/O device for inputting and outputting data. The data bus provides a path over which data is transferred between the microprocessor, memory, and I/O interface.

There are two basic trends in the physical make-up of microcomputers. One is a multichip approach where the microprocessor, memory, and I/O interface units are housed separately. The other is a single-chip approach where the microprocessor, memory, and I/O are integrated on a single chip. Therefore, the microprocessor is simply the processing unit of the microcomputer and has no memory or I/O interface of its own.

ELEMENTS OF A MICROPROCESSOR

Before delving into the components of a microprocessing unit, here is a brief review of the concepts of words and bytes.

A complete unit of binary information or data is called a *word*. For instance, the number 200_{10} can be represented by eight bits as 11001000_2. This, then, is an eight-bit word because it completely represents the decimal number 200. Now consider the number 32,768. This number cannot be represented with eight bits but requires 16 bits—1000000000000000_2. Hence, a 16-bit word is required for expressing all decimal numbers from 0 to 65,535.

Most microprocessors handle bits in eight-bit groups called bytes. A byte can be a word or only part of a word. For instance, the eight-bit word mentioned above consists of one byte, however, the 16-bit word consists of two bytes.

Figure 13-3 shows a block diagram of a microprocessor. Many of the elements contained in this microprocessor are common to most other microprocessors, although the internal arrangement of *architecture* normally varies from one manufacturer's device to another's.

Arithmetic Logic Unit (ALU)

The arithmetic logic unit contains the logic that performs all arithmetic and logic operations. Data is brought into the ALU from one or both of the registers called *accumulators* and from the *data register*. Figure 13-3 shows two accumulators. Both the accumulators and the data register are eight-bit registers that store one byte of data. Each byte brought into the ALU is called an *operand*. This is the group of bits that the ALU will be manipulating.

For example, FIG. 13-4 shows an eight-bit number from accumulator A being added to an eight-bit number from the data register. The result of this addition (sum) is put into an accumulator, replacing the original operand that was stored there. When the ALU performs an operation on two operands, the result of the operation always goes into an accumulator and replaces the previous operand. Therefore, the accumulator not only stores an operand prior to manipulation by the ALU, but it also stores the result

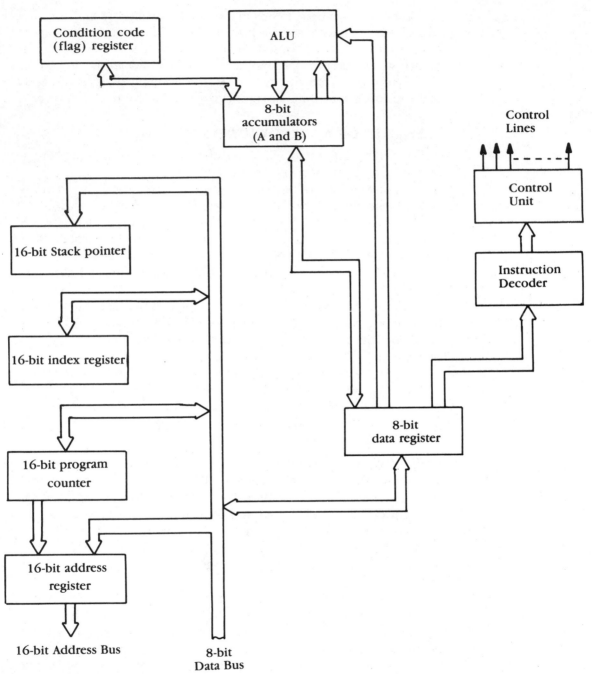

13-3 Simplified microprocessor block diagram.

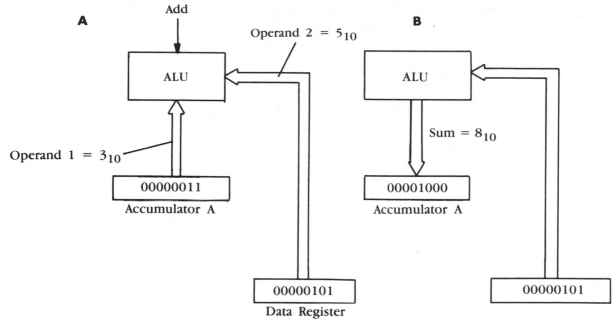

13-4 Example of an ALU adding two operands. (A) is the ALU adding 011_2 and 101_2. (B) shows the sum (1000_2) being put into the accumulator.

of the operation after the operation has been performed. The data register temporarily stores a byte of data that is to be put onto the data bus or has been taken off the data bus.

Instruction Decoder and Control Unit

An instruction is a binary code that tells the microprocessor what it is to do. An orderly arrangement of many different instructions makes up a typical *program*. A program is a step-by-step procedure used by the microprocessor to carry out a specific task.

The instruction decoder within the microprocessor decodes an instruction code that has been transferred onto the data bus from the memory. The instruction code is commonly called an operation code or *op code*. When the op code is decoded, the instruction decoder provides the control unit with this information so it can produce the proper signal and timing sequence to execute the instruction.

Condition Code Register

This register is sometimes called a *flag* register or *status* register. Its basic purpose is to indicate the status of the contents of the accumulator or certain other conditions within the microprocessor. For example, it can indicate a zero result, a negative result, the occurrence of a carry, and the occurrence of an overflow from the accumulator.

Program Counter

This is usually a 16-bit counter that produces the sequence of memory addresses from which the program instructions are taken. The content of the program counter is always the memory address from which the next byte is to be taken. In some microprocessors, the program counter is known as the *instruction pointer.*

Address Register

This is usually a 16-bit register that temporarily stores an address from the program counter in order to put it onto the address bus. As soon as the program counter loads an address into the address register, it is incremented (increased by one) to the address of the next instruction.

Stack Pointer and Index Register

The *stack pointer* is usually a 16-bit register, used mainly during subroutines and interrupts (described later in this chapter). It is used in conjunction with the memory stack (discussed in chapter 10). The *index register* is also usually a 16-bit register and is used as one means of addressing the memory. It is used with a mode of addressing called *indexed addressing.*

Address Bus and Data Bus

The address bus usually consists of 16 parallel lines to accommodate a 16-bit address code. This allows the microprocessor to address up to $2^{16} = 65,536$ bytes of memory. Some microprocessors have more address bits so that larger memories can be addressed.

The data bus commonly consists of eight parallel lines (hence, it would be an *eight-bit microprocessor*). One byte of data can be transferred to or from the memory of I/O on this bus at any given time. Some μPs have a 16-bit data bus (*16-bit microprocessors*). The remainder of the examples in this chapter refer to an eight-bit device.

THE MICROPROCESSOR AND THE MEMORY

As mentioned, the microprocessor is connected to a memory with the address bus and the data bus. In addition, there are certain control signals that must be sent between the microprocessor and the memory such as the *read* and *write* controls. This is illustrated in FIG. 13-5.

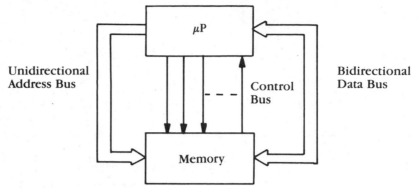

13-5 Microprocessor/memory interaction.

As indicated, the address bus is unidirectional, meaning the address data bits go only one way—from the microprocessor to the memory. (There is no need for the memory to send any address information to the µP). The data bus is bidirectional so information can travel either to or from the microprocessor or memory.

Read Operation

To transfer a byte of data from the memory to the microprocessor, a *read* operation must be carried out. Refer to the example in FIG. 13 6. To begin, the program counter contains the 16-bit address of the byte to be read from the memory. This address is loaded into the address register and put onto the address bus. The program counter advances by one (increments) to the next address and waits.

Once the address code is on the bus, the microprocessor control unit sends a read signal to the memory. At the memory, the address bits are decoded and the desired memory location is selected. The read signal causes the *contents* of the selected address to be put on the data bus. The data byte is then loaded into the data register to be used by the µP. This completes the read operation.

Note that each memory location contains one byte of data. When a byte is read from memory, it is not destroyed but remains in the memory. This process of "copying" the contents of a memory location without destroying its contents is called *nondestructive readout.*

Write Operation

In order to transfer a byte of data from the µP to the memory, a *write* operation is required. This is illustrated in FIG. 13-7. The memory is addressed in the same way as the read operation. A data byte being held in the data register is put onto the data bus and the microprocessor sends

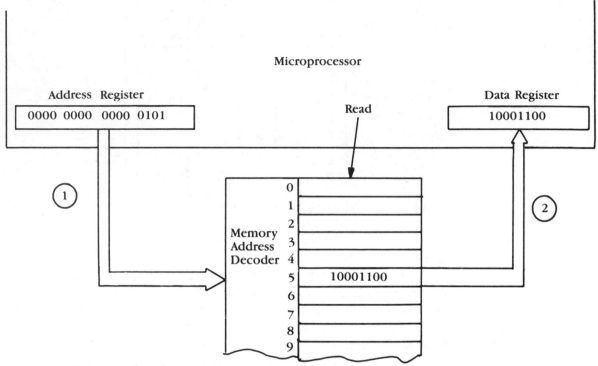

13-6   Read operation. ① shows address 5_{10} being put on the address but followed by a *read* signal. ② shows the contents of address 5_{10} in memory put on the data bus and stored in a data register.

the memory a *write* signal. This causes the byte on the data bus to be stored at the selected location in the memory as specified by the 16-bit address code. The existing contents of that particular memory location are replaced by the new data byte. This completes the write operation.

HEXADECIMAL REPRESENTATION OF ADDRESS AND DATA

The only characters a microprocessor recognizes are 1's and 0's. However, most literature on microprocessors uses the hexadecimal number system to simplify the representation of binary quantities for programmers.

For instance, the binary address 0000000000001111 can be written as 000F in hexadecimal. A 16-bit address can have a minimum hexadecimal value of 0000_{16} and a maximum value of $FFFF_{16}$. With this notation, a 64K memory (actually 65,536) can be shown in block form as in FIG. 13-8. (The hexadecimal code is listed in TABLE 2-2.)

A data byte can also be represented in hexadecimal. A data byte can be either an instruction, an operand, or an address. A data byte is eight bits and can represent decimal numbers from 0_{10} to 255_{10}, or it can repre-

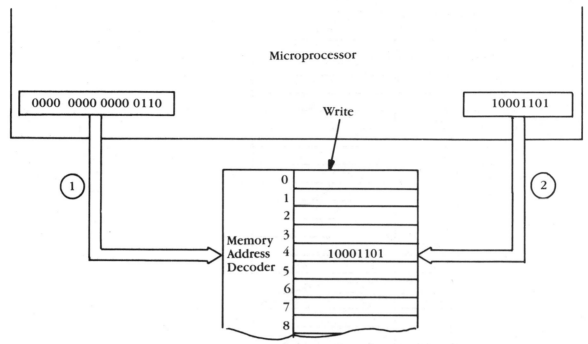

13-7   Write operation. ① represents the address on the address bus and ② is data on the data bus. The *write* signal places it in the address location 6_{16}.

sent up to 256 instructions. For example, a microprocessor code that is 10001100 in binary is written as 8C in hexadecimal.

FETCH AND EXECUTE

When a program is running, the microprocessor goes through a repetitive sequence consisting of two fundamental phases, as shown in FIG. 13-9. One phase is called *fetch* and the other is *execute*. During the fetch phase, an instruction is read from the memory and decoded by the instruction decoder. During the execute phase, the μP carries out the sequence of operations indicated by that particular instruction. As soon as one instruction has been executed, the microprocessor returns to the fetch phase to get the next instruction from the memory. Fetch/execute is demonstrated in the section discussing specific instructions later in this chapter.

INSTRUCTION FORMAT

A microprocessor must address the memory to obtain or store data. There are several ways in which to generate an address when an instruction is

Address (hexidecimal)	Contents
0000	
0001	
0002	
0003	
0004	
0005	
0006	
0007	
0008	
0009	
000A	
000B	
000C	
FFFB	
FFFC	
FFFD	
FFFE	
FFFF	

I3-8 Representation of a 64K memory.

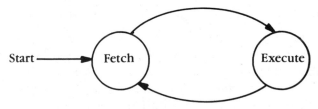

I3-9 Fetch and execute.

being executed. These are called the *addressing modes* of the microprocessor, and they provide for wide programming flexibility.

The 6800 has several addressing modes. Each of the instructions
available to the microprocessor has a certain addressing mode associated
with it. These are inherent, immediate, direct, extended, relative, and indexed. (The addressing modes are not covered in this text; their details
are outside the scope of this introduction.)

Instructions

The instructions for this section are those of the Motorola 6800 microprocessor. This unit is a good representation of a basic microprocessor from which to learn. Microprocessors typically have at least 50 instructions that make up what is called the *instruction set*. The 6800 has 72 instructions. Although instruction sets vary from one manufacturer to another, many instructions are common to all µPs.

The following instructions are demonstrated in this section: load accumulator (LDAA), store accumulator (STAA), addition (ADDA), and clear accumulator (CLRA). The A on the end of each of these instructions indicates that they apply to accumulator A (the 6800 has two accumulators). There are similar instructions for operating with accumulator B. The abbreviated designations for the instructions—LDAA, STAA, ADDA, and CLRA—are called *mnemonics*.

Format

A microprocessor instruction can consist of one, two, or three bytes, depending on the instruction. A one-byte instruction is illustrated in FIG. 13-10A. This type of instruction requires one memory location. The eight-bit instruction code (the op code) uniquely identifies that instruction.

A two-byte instruction is illustrated in FIG. 13-10B. The first byte is the op code, and the second byte can be either an operand or a code associated with a memory address. This type of instruction is stored in two consecutive memory locations. Finally, FIG. 13-10C shows a three-byte in-

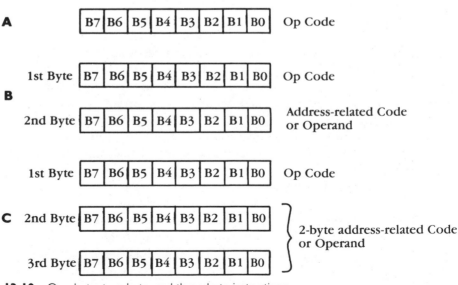

13-10 One-byte, two-byte, and three-byte instructions.

struction. In this case, the first byte is the op code and the second and third bytes can be an operand or an address-associated code. This type of instruction is stored in three consecutive memory locations.

A SIMPLE PROGRAM

In this section, a very simple program is used to illustrate how a program is run on the microprocessor. This program adds two numbers and stores the sum in the memory. Initially, the program is stored at memory addresses 0000_{16} through 0006_{16}.

Figure 13-11 shows the program using the *load* (LDAA) and *add*

Memory Address	Memory Contents		Mnemonic/ Contents
	Binary	Hexadecimal	
0000_{16}	10000110	86	LDAA
0001_{16}	00001000	08	8_{10}
0002_{16}	10001011	8B	ADDA
0003_{16}	00001100	OC	12_{10}
0004_{16}	10010111	97	STAA
0005_{16}	00000111	07	7_{10}
0006_{16}	00111110	3E	WAI
0007_{16}			Reserved for sum

13-11   A program to add 8_{10} and 12_{10} and store the sum in memory.

(ADDA) instructions. The two numbers to be added are 8_{10} and 12_{10}. The first operand (8_{10}) follows the LDAA instruction in memory, and the second operation (12_{10}) follows the ADDA instruction. The *store* (STAA) instruction stores the sum of the two operands at memory address 0007_{16}.

The sequence of operations is as follows. Figure 13-12 shows each step. In part (A), the μP fetches the LDAA instruction from memory by performing a read operation at address 0000_{16}. The program counter is then advanced to 0001_{16}.

Part (B) illustrates the execution of the LDAA instruction. The operand 8_{10} is read from address 0001_{16} and loaded into the accumulator. The program counter then advances to 0002_{16}.

Part (C) shows the μP fetching the ADDA instruction from memory address 0002_{16}. The program counter then advances to 0003_{16}.

Part (D) shows the execution of the ADDA instruction. This involves reading the operand 12_{10} from address 0003_{16} and then adding it to the first operand (8_{10}), which is stored in the accumulator. The sum of these two numbers is then loaded into the accumulator, replacing the first operand. The program counter then advances to 0004_{16}.

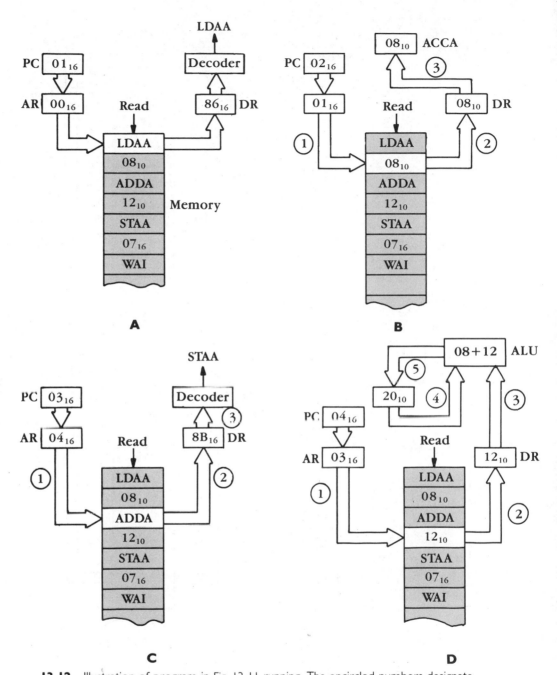

13-12 Illustration of program in Fig. 13-11 running. The encircled numbers designate the sequence of events within the step.
(A) Fetch LDAA op code. (B) Load operand 08_{10} into accumulator A. (C) Fetch ADDA op code. (D) Operand 12_{10} is read and added to 8_{10}. The sum 20_{10} is placed in accumulator A. (E) Fetch STAA op code. (F) Address 06_{16} is read and transferred into address register. (G) Sum is stored at address 07_{16}. (H) Fetch WAI op code.

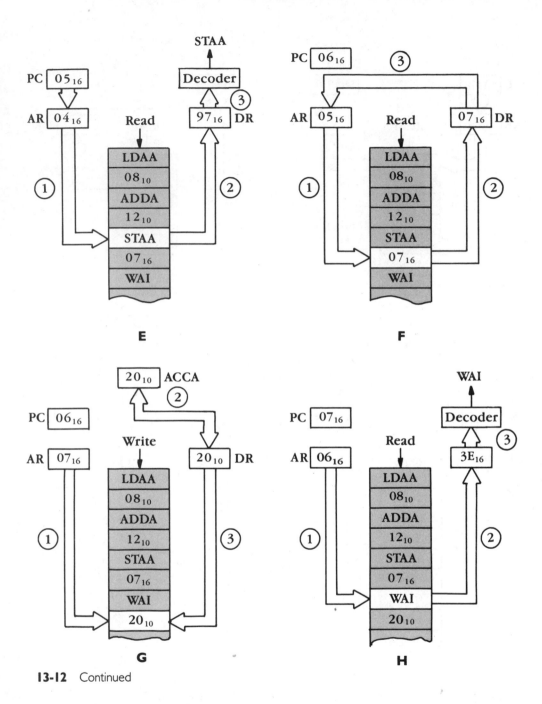

13-12 Continued

Next, the microprocessor fetches the STAA instruction as shown in part (E), and the program counter advances to 0005_{16}. Parts (F) and (G) show the execution of STAA in which 07_{16} is read from memory address 0005_{16} and loaded into the address register. The sum in the accumulator is then stored at address 07_{16}. The program counter than advances to 0006_{16}.

Part (H) shows the fetch and execution of the *wait* instruction (WAI), which ends the program.

SUBROUTINES

A subroutine is basically a program within a main program. A subroutine is normally used when a specific operation is "called" repeatedly during the running of the main program. The 6800 μP has several instructions that can be used for subroutines. These include JMP (jump), JSR (jump to sub-routine), and RTS (return from subroutine).

JMP Instruction

This instruction allows the μP to jump from one point in the program to another. This is a three-byte instruction, as shown in FIG. 13-13. The first byte is the op code ($7E_{16}$). The second and third bytes are the MSB and LSB, respectively, of the address to which the microprocessor jumps.

JMP Op Code ($7E_{16}$)

Designation Address {

Most Significant Byte

Least Significant Byte

13-13   JMP instruction format.

As an example, FIG. 13-14 shows the JMP instruction as part of the main program. This causes a jump to address $B000_{16}$, which is the beginning of the subroutine. At the end of the subroutine, another JMP causes the μP to jump back to where it left off in the main program.

The JMP instruction is useful where a program is to be repeated in an endless loop. A limitation of the JMP instruction is that the μP can return to only *one* address in the main program. So a jump from several points in the main program to a single subroutine and back to the appropriate address is not feasible. The JSR and RTS instructions overcome this limitation.

JSR and RTS Instructions

The JSR is a three-byte instruction with a format similar to JMP. RTS is a single-byte instruction. The second and third bytes of the JSR instruction specify the beginning address of the subroutine to which the program is to jump.

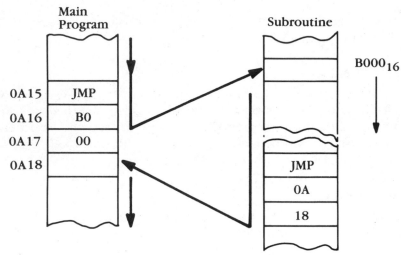

13-14   A jump to a subroutine and back.

When JSR is executed, the program counter contents are pushed onto the memory stack. At the end of the subroutine, the RTS instruction causes the program counter to be pulled from the stack, which causes the μP to return to the address in the main program where it left off. This allows the subroutine to be called several times during a main program, where each time a different return address is required. Figure 13-15 illustrates the use of JSR and RTS instructions for calling a subroutine three times during the main program.

INPUT/OUTPUT OPERATIONS

To be useful, the microprocessor system must be able to accept data from external devices, such as a keyboard, and to send results to external devices, such as video terminals or printers. In the 6800 microprocessor, input and output (I/O) operations are handled as transfers to or from a memory address. That is, an I/O device is assigned an address and treated as a memory location.

For example, a keyboard can be assigned a two-byte address. To transfer a byte of data representing a keyboard character into the accumulator, the microprocessor executes an LDAA instruction. The second two bytes of this instruction make up the address assigned to the keyboard.

When the microprocessor puts this address on the address bus, a keyboard decoder enables the data from the keyboard to load onto the data bus to be transferred into the microprocessor system. Figure 13-16 illustrates this operation for a keyboard address of $C000_{16}$.

An output device is also assigned a memory address. To transfer a

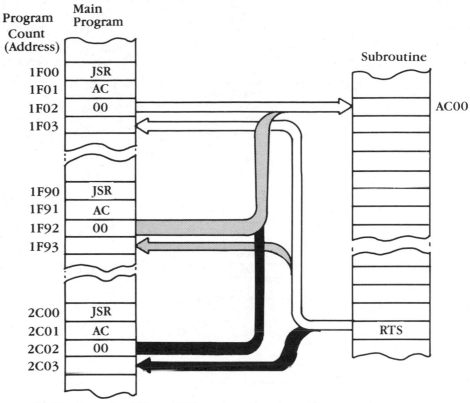

13-15 Example of JSR and RTS used to call a subroutine several times.

byte of data from accumulator A to the external device, the μP executes a STAA instruction. The second two bytes of this instruction make up the address assigned to the device.

When the STAA instruction is executed, the output device address is placed on the address bus and the data byte on the data bus. The device decoder responds to the address code by enabling the output buffers, thus effecting the transfer to the device. For example, FIG. 13-17 shows an output operation in which the external device is a printer assigned an address of $D000_{16}$. When the address code is placed on the address bus, the printer's decoder enables the data transfer.

Hence, the 6800 handles input/output operations as data transfers to and from memory using the STAA and the LDAA instructions. These are programmed or software I/O operations, and the method is called *memory-mapped I/O.* Other microprocessors often have special input and output instructions rather than treating I/O operations as memory transfers.

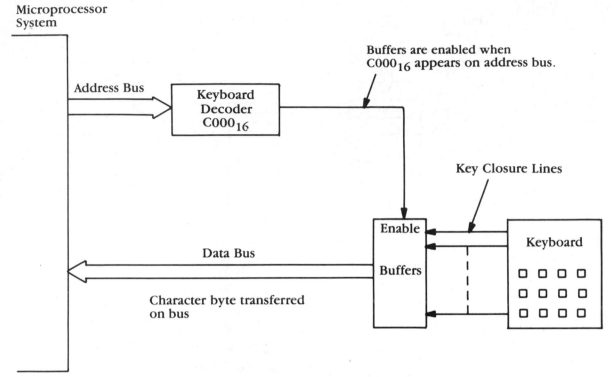

13-16   Example of input of data byte from keyboard.

Interrupts

Another way of handling I/O transfers uses a concept known as *interrupts*. Rather than using a program-controlled I/O transfer as discussed in the previous section, the interrupt allows the external device to tell the microprocessor that it is ready to send data or ask the microprocessor to send data. Hence, an interrupt allows the external device to interrupt the microprocessor's normal operations to request service.

The idea behind interrupts is that when any device requests some action, the microprocessor must know what to do to service the device. To accomplish this, the interrupt commands are the beginning address of the service routine. When the microprocessor gets a particular type of interrupt, it completes its current instruction and goes to the address in ROM where the interrupt directive is. This interrupt byte(s) tells the microprocessor where to go in RAM to begin the sequence required to service the interrupt device.

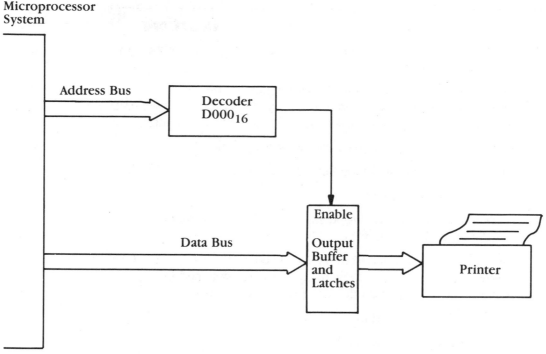

13-17 Example of output of data byte to printer. The enable signal from the decoder to the output buffer allows the data to be transferred to the printer.

Summary

The microprocessor is one of the main components of the microcomputer, together with memory, input/output devices, an input/output interface, and buses. The microprocessor consists of a group of registers, together with an ALU. Some of the registers are an accumulator, an address register, and data registers. Information in these registers is normally stored in 8-bit blocks, called *bytes*. Program instructions are brought from memory to an instruction decoder, which causes input/output and arithmetic operations to occur. The two primary memory operations are *read* and *write*. Information on the address bus indicates the memory locations involved, and information on the data bus may be data, instructions, or addresses. Both memory locations and input/output devices have specific addresses, so that data can be directed properly.

Questions

1. Draw a block diagram of a microcomputer, labeling each of the parts and stating its function.

2. State the function of each of the following parts of the microprocessor: address register, index register, arithmetic logic unit, accumulator, instruction decoder, control unit, and stack pointer.

3. What are the two different types of memory operations?

4. Why must input/output devices have addresses?

5. What is an *instruction set?*

6. If two 8-bit words are used to identify an address, how many different locations are there?

7. In FIG. 13-6, the address register is 0000 0000 0000 0101, and the data register is 1000 1100.   **a.** How many bytes are contained in each register?   **b.** What is the value of each in hexadecimal?

8. Why is a LIFO stack used in subroutines?

9. What are *interrupts?* Why are they used?

APPENDIX A
COMMON
ABBREVIATIONS

ASCII	American Standard Code for Information Interchange
BCD	*b*inary-*c*oded *d*ecimal
Bit	*b*inary dig*it*
CAM	*c*ontent-*a*ddressable *m*emory
CML	*c*urrent-*m*ode *l*ogic
CMOS	*c*omplementary *m*etal-*o*xide *s*emiconductor
DCTL	*d*irect-*c*oupled *t*ransistor *l*ogic
DEMUX	*demu*ltiple*x*
DTL	*d*iode-*t*ransistor *l*ogic
ECL	*e*mitter-*c*oupled *l*ogic
FET	*f*ield-*e*ffect *t*ransistor
FIFO	*f*irst *i*n-*f*irst *o*ut
IC	*i*ntegrated *c*ircuit
LED	*l*ight-*e*mitting *d*iode
LIFO	*l*ast *i*n-*f*irst *o*ut
LSB	*l*east *s*ignificant *b*it
LSD	*l*east *s*ignificant *d*igit
LSI	*l*arge-*s*cale *i*ntegration
MBM	*m*agnetic *b*ubble *m*emory
μP	microprocessor
MNOS	*m*etal-*o*xide *n*itride *s*emiconductor
MOS	*m*etal-*o*xide *s*emiconductor
MSB	*m*ost *s*ignificant *b*it
MSD	*m*ost *s*ignificant *d*igit
MSI	*m*edium-*s*cale *i*ntegration
MUX	*mu*ltiple*x*

PLA	*programmable logic array*
PROM	*programmable read-only memory*
RAM	*random-access memory*
RCTL	*resistor-capacitor-transistor logic*
RMM	*read-mostly memory*
ROM	*read-only memory*
RTL	*resistor-transistor logic*
TTL	*transistor-transistor logic*
UART	*universal asynchronous receiver transmitter*

APPENDIX B
POWERS
OF NUMBERS

n	n^4	n^5	n^6	n^7	n^8
1	1	1	1	1	1
2	16	32	64	128	256
3	81	243	729	2187	6561
4	256	1024	4096	16384	65536
5	625	3125	15625	78125	390625
6	1296	7776	46656	279936	1679616
7	2401	16807	117649	823543	5764801
8	4096	32768	262144	2097152	16777216
9	6561	59049	531441	4782969	43046721
					$\times 10^8$
10	10000	100000	1000000	10000000	1.000000
11	14641	161051	1771561	19487171	2.143589
12	20736	248832	2985984	35831808	4.299817
13	28561	371293	4826809	62748517	8.157307
14	38416	537824	7529536	105413504	14.757891
15	50625	759375	11390625	170859375	25.628906
16	65536	1048576	16777216	268435456	42.949673
17	83521	1419857	24137569	410338673	69.757574
18	104975	1889568	34012224	612220032	110.119606
19	130321	2476099	47045881	893871739	169.835630
				$\times 10^9$	$\times 10^{10}$
20	160000	3200000	64000000	1.280000	2.560000
21	194481	4084101	85766121	1.801089	3.782286
22	234256	5153632	113379904	2.494358	5.487587
23	279841	6436343	148035889	3.404825	7.831099

n	n^4	n^5	n^6	n^7	n^8
24	331776	7962624	191102975	4.586471	11.007531
25	390625	9765625	244140625	6.103516	15.258789
26	456976	11881376	308915776	8.031810	20.882706
27	531441	1438907	387420489	10.460353	28.242954
28	614656	17210368	481890304	13.492929	37.780200
29	707281	20511149	594823321	17.249876	50.024641
			$\times 10^8$	$\times 10^{10}$	$\times 10^{11}$
30	810000	24300000	7.290000	2.187000	6.561000
31	923521	28629151	8.875037	2.751261	8.528910
32	1048576	33554432	10.737418	3.435974	10.995116
33	1185921	39135393	12.914680	4.261844	14.064086
34	1336336	45435424	14.448044	5.252335	17.857939
35	1500625	52521875	18.382656	6.433930	22.518754
36	1679616	60466176	21.767823	7.836416	28.211099
37	1874161	69343957	25.657264	9.493188	35.124795
38	2085136	79235168	30.109364	11.441558	43.477921
39	2313441	90224199	35.187438	13.723101	53.520093
		$\times 10^9$	$\times 10^{10}$	$\times 10^{12}$
40	2560000	102400000	4.096000	16.384000	6.553600
41	2825761	115856201	4.750104	19.475427	7.984925
42	3111696	130691232	5.489032	23.053933	9.682652
43	3418801	147008443	6.321363	27.181861	11.688200
44	3748096	164916224	7.256314	31.927781	14.048224
45	4100625	184528125	8.303766	37.366945	16.815125
46	4477456	205962976	9.474297	43.581766	20.047612
47	4879681	229345007	10.779215	50.662312	23.811287
48	5308416	254803968	12.230590	58.706834	28.179280
49	5764801	282475249	13.841287	67.822307	33.232931
			$\times 10^9$	$\times 10^{11}$	$\times 10^{13}$
50	6250000	312500000	15.625000	7.8125000	3.9062500
51	6765201	345025251	17.596288	8.974107	4.576794
52	7311616	380204032	19.770610	10.280717	5.345973
53	7890481	418195493	22.164361	11.747111	6.225969
54	8503056	459165024	24.794911	13.389252	7.230196
55	9150625	503284375	27.680641	15.224352	8.373394
56	9834496	550731776	30.840979	17.270948	9.671731
57	10556001	601692057	34.296447	19.548975	11.142916
58	11316496	656356768	38.068693	22.079842	12.806308
59	12117361	714924299	42.180534	24.886515	14.683044
		$\times 10^8$	$\times 10^{10}$	$\times 10^{11}$	$\times 10^{13}$
60	12960000	7.776000	4.665600	27.993600	16.796160
61	13845841	8.445963	5.152037	31.427428	19.170731

n	n^4	n^5	n^6	n^7	n^8
62	14776336	9.161328	5.680024	35.216146	21.834011
63	15752961	9.924365	6.252350	39.389806	24.815578
64	16777216	10.737418	6.871948	43.980465	28.147498
65	17850625	11.602906	7.541889	49.022279	31.864481
66	18974736	12.523326	8.265395	54.551607	36.004061
67	20151121	13.501251	9.045838	60.607116	40.606768
68	21381376	14.539336	9.886748	67.229888	45.716324
69	22667121	15.640313	10.791816	74.463533	51.379837
		$\times 10^8$	$\times 10^{10}$	$\times 10^{12}$	$\times 10^{14}$
70	24010000	16.807000	11.764900	8.235430	5.764801
71	25411681	18.042294	12.810028	9.095120	6.457535
72	26873856	19.349176	13.931407	10.030613	7.222041
73	28398241	20.730716	15.133423	11.047399	8.064601
74	29986576	22.190066	16.420649	12.151280	8.991947
75	31640625	23.730469	17.797852	13.348389	10.011292
76	33362176	25.355254	19.269993	14.645195	11.130348
77	35153041	27.067842	20.842238	16.048523	12.357363
78	37015056	28.871744	22.519960	17.565569	13.701144
79	38950081	30.770564	24.308746	19.203909	15.171088
		$\times 10^8$	$\times 10^{10}$	$\times 10^{12}$	$\times 10^{14}$
80	40960000	32.768000	26.214400	20.971520	16.777216
81	43046721	34.867844	28.242954	22.876792	18.530202
82	45212176	37.073984	30.400667	24.928547	20.441409
83	47458321	39.390406	32.694037	27.136051	22.522922
84	49787136	41.821194	35.129803	29.509035	24.787589
85	52200625	44.370531	37.714952	32.057709	27.249053
86	54700816	47.042702	40.456724	34.792782	29.921893
87	57289761	49.842092	43.362620	37.725479	32.821167
88	59969536	52.773192	46.440409	40.867560	35.963452
89	62742241	55.840594	49.698129	44.231335	39.365888
		$\times 10^9$	$\times 10^{11}$	$\times 10^{13}$	$\times 10^{15}$
90	65610000	5.904900	5.314410	4.782969	4.304672
91	68574961	6.240321	5.678693	5.167610	4.702525
92	71639296	6.590815	6.063550	5.578466	5.132189
93	74805201	6.956884	6.469902	6.017009	5.595818
94	78074896	7.339040	6.898698	6.484776	6.095689
95	81450625	7.737809	7.350919	6.983373	6.634204
96	84934656	8.153727	7.827578	7.514475	7.213896
97	88529281	8.587340	8.329720	8.079828	7.837434
98	92236816	9.039208	8.858424	8.681255	8.507630
99	96059601	9.509900	9.414801	9.320653	9.227447
100	100000000	10.000000	10.000000	10.000000	10.000000

ANSWERS TO ODD-NUMBERED PROBLEMS

Chapter 1

1. a. 13, 15   **b.** 9.75, 11.6
3. a. 11101, 35   **b.** 110011.5, 63.4
5. a. 100011, 35   **b.** 111011.101, 59.675
7. a. 0 0011, 0 0011, 0 0011   **b.** 1 1111, 1 0000, 1 0001

Chapter 2

1. a. 10000111   **b.** 0001 0011 0101   **c.** 0100 0110 1000   **d.** 87
3. 11100010
5. 1
7. a. 110   **b.** 1100110   **c.** Check parity bits for received code. Parity bits are 111 (P3P2P1). This means that the error is in the seventh bit from the left.
9. 100 0101   100 1110   100 0100

Chapter 3

1. a. 11110   **b.** 10100   **c.** 1111101   **d.** 101
3. a. 0 10010   **b.** 0 100100   **c.** 1 11110011   **d.** 1 10
5. a. 1 1100   **b.** 0 10101
7. a. 1 1100   **b.** 0 10110
9. a. 1111 0101   **b.** 101
11. 1000 0001 0011
13. a. J = 1000 1100 1010, K = 0101 0100 1001
    b. 1101 10000 10011
    c. 1011 0100 0110
15. a. J = 1000 1100 1010, K = 0101 0100 1001   **b.** 0011 1000 0001
    c. 0110 1011 0100

Chapter 4

1. a.

A	B
0	0
1	1

b.

A	B	F
0	0	0
0	1	1
1	0	1
1	1	1

c.

A	B	F
0	0	0
0	1	0
1	0	0
1	1	1

d.

A	B	F
0	0	0
0	1	1
1	0	1
1	1	0

3. a.

A ———▷——— B

b.

A, B ——⟩——— F

c.

A, B ——⟩——— F

d.

A, B ——⟩⟩——— F

Chapter 5

1.

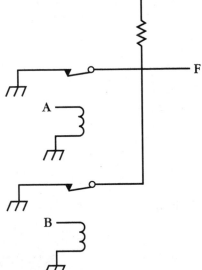

3.

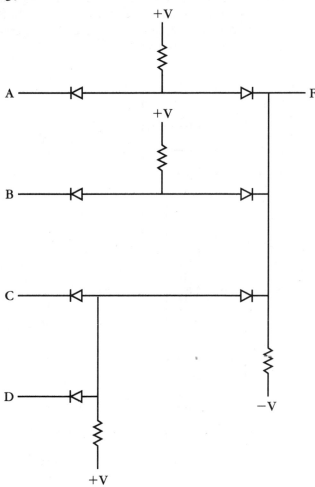

5.

7. A, B, low: base of transistor low, transistor is off, F at +V; A high, B low: base of transistor high, transistor is on, F is low; A low, B high: base of transistor high, transistor is on, F is low; A and B high: base of transistor high, transistor is on, F is low

9.

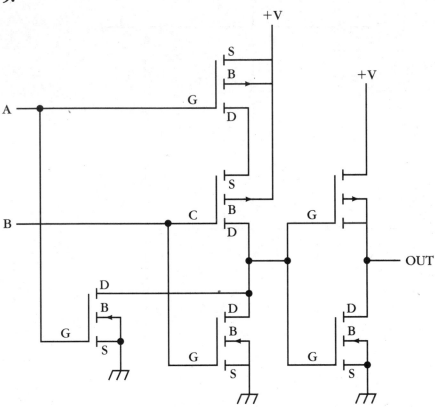

Chapter 6

1. $A + C = F$

3. a. $A + BC = F$ **b.** $AB + \overline{A}\overline{B} = F$

5. a.

	B 0	1
A 0	1	0
1	0	1

b.

AB	C 0	1
00	1	1
01	1	1
11	0	0
10	1	0

7.

AB \ C	0	1
00	1	0
01	0	1
11	0	0
10	1	1

$A\overline{B} + \overline{B}\,\overline{C} + \overline{A}BC = F$

Chapter 7

1. a. $\overline{\overline{\overline{AB}}C} = X$ **b.**

C	B	A	$\overline{A}$	$\overline{B}$	$\overline{\overline{AB}}$	$\overline{\overline{AB}}C$	X
0	0	0	1	1	0	0	1
0	0	1	0	1	1	0	1
0	1	0	1	0	1	0	1
0	1	1	0	0	1	0	1
1	0	0	1	1	0	0	1
1	0	1	1	0	1	1	0
1	1	0	0	1	1	1	0
1	1	1	0	0	1	1	0

3.

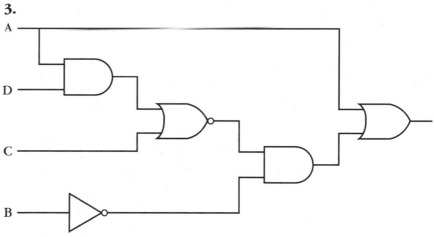

5. a. 4   **b.** $A + \overline{B}\,\overline{C}$ **c.**

AB \ CD	00	01	11	10
00	1	1	0	0
01	0	0	0	0
11	1	1	1	1
10	1	1	1	1

7. One of the inputs to gate G4 is shorted. Connect *C* to +V; if output is still high, other input is shorted.

9.

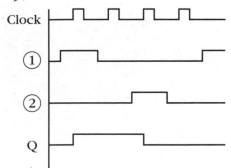

11.

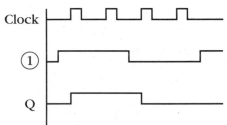

Chapter 8

1. Clear should go high to clear and then should stay low. Preset then goes high, allowing data in and placing bit D on the output. Three clock pulses later, the entire 4-bit word has been clocked out.

3.

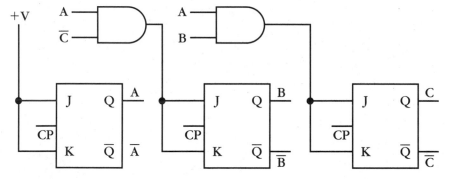

Chapter 9

1. Remove the QB input to the AND gate.
3. 16

5. D2

7. The inverted output is the parity bit.

9.

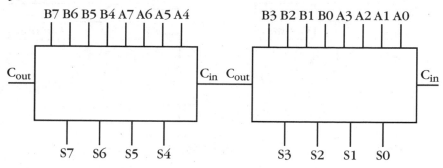

Chapter 10

1.

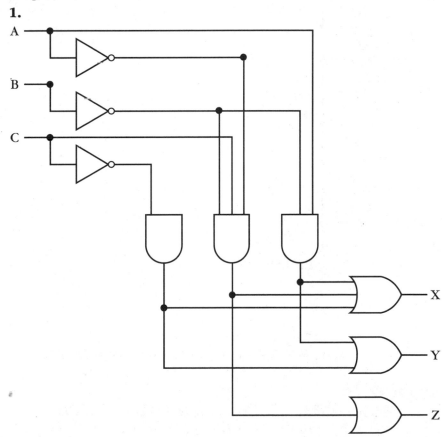

3.

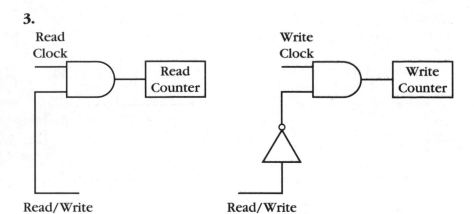

Read/Write   Read/Write

Chapter 11

1.

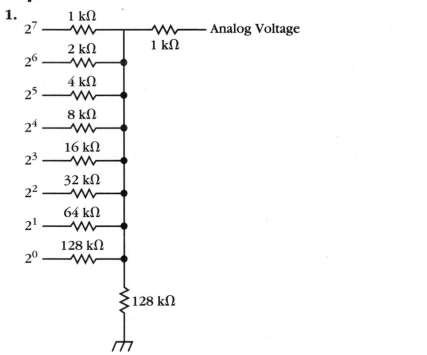

3. a. 11 V   **b.** −12.5 V   **c.** −10 V

5. LSB = 0.079 V, 0.159 V, 0.317 V, 0.635 V, 1.270 V; MSB= 2.540 V

Chapter 12

1. Segment A: $\overline{A}\,\overline{B}\,\overline{C}\,\overline{D} + \overline{A}BC\overline{D} + AB\overline{C}\,\overline{D} + \overline{A}\,\overline{B}C\overline{D} + A\overline{B}\,\overline{C}D$
$+ ABC\overline{D} + A\overline{B}C\overline{D} = AC + \overline{A}\,\overline{C} + B\overline{C} + D$

Segment B: $\overline{A}\overline{B}\overline{C}\overline{D} + A\overline{B}\overline{C}\overline{D} + \overline{A}B\overline{C}\overline{D} + AB\overline{C}\overline{D} + \overline{A}\overline{B}C\overline{D} +$
$AB C\overline{D} + \overline{A}\overline{B}\overline{C}D + A\overline{B}\overline{C}D = AB + \overline{A}\overline{B} + \overline{C}$

Segment C: $\overline{A}\overline{B}\overline{C}\overline{D} + A\overline{B}\overline{C}\overline{D} + AB\overline{C}\overline{D} + \overline{A}\overline{B}C\overline{D} + A\overline{B}C\overline{D} +$
$\overline{A}BC\overline{D} + ABC\overline{D} + \overline{A}\overline{B}\overline{C}D + A\overline{B}\overline{C}D = A + \overline{B} + C$

Segment D: $\overline{A}\overline{B}\overline{C}\overline{D} + \overline{A}\overline{B}\overline{C}D + A\overline{B}\overline{C}D + AB\overline{C}D + \overline{A}BC\overline{D} +$
$A\overline{B}C\overline{D} = \overline{A}B + \overline{A}\overline{C} + B\overline{C} + A\overline{B}C$

INDEX